OXFORD BIOLOGY PRIMERS

'The future of biological research rests in the minds of the current generation of students. The Oxford Biology Primers will play an important role in inspiring those students to want to understand more about the world around them, in so doing setting them on the path of exploration and enquiry.'

Sir Paul Nurse Hon FRSB, Director of the Crick Institute

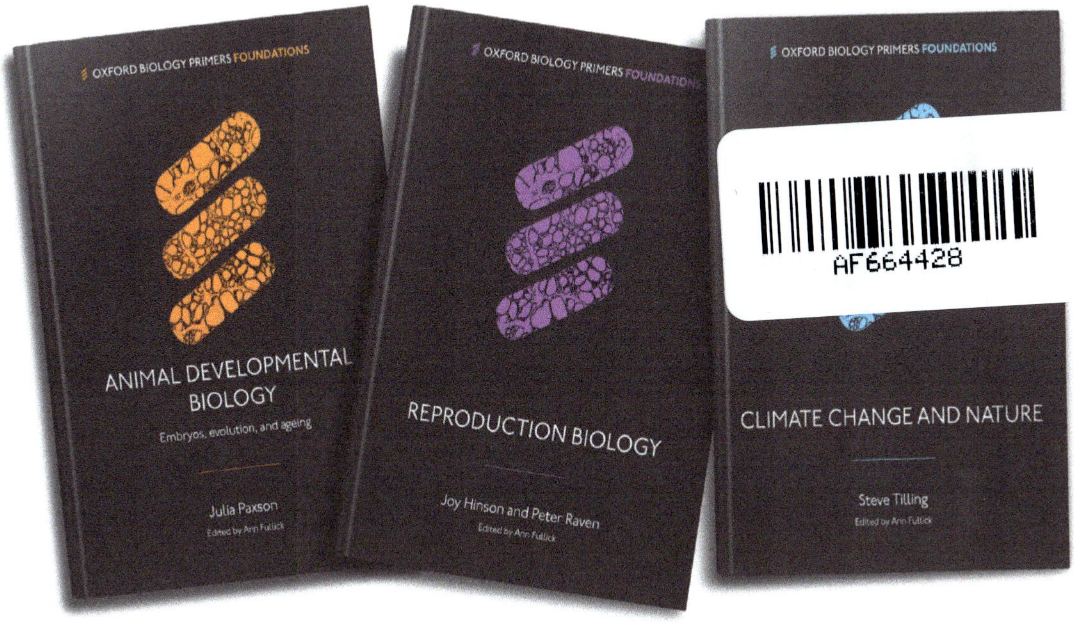

'The Royal Society of Biology is delighted to have helped develop this invaluable series of biology textbooks. The only resource of its kind, the series brings together foundational biological theory with research from the frontiers of modern bioscience, on topics spanning the full breadth of life on Earth.'

Professor Dame Julia Goodfellow CBE CBiol FRSB, RSB President

Discover more in the series at

www.oxfordtextbooks.co.uk/obp

Published in partnership with the Royal Society of Biology

Royal Society of **Biology**

EXPERIMENTATION AND STATISTICS FOR BIOLOGISTS

OXFORD BIOLOGY PRIMERS

EXPERIMENTATION AND STATISTICS FOR BIOLOGISTS

Stephen C. Land
University of Dundee

Ann Fullick
Series Editor

OXFORD UNIVERSITY PRESS

Royal Society of Biology

OXFORD
UNIVERSITY PRESS

Great Clarendon Street, Oxford, OX2 6DP,
United Kingdom

Oxford University Press is a department of the University of Oxford. It furthers the University's objective of excellence in research, scholarship, and education by publishing worldwide. Oxford is a registered trade mark of Oxford University Press in the UK and in certain other countries.

Published in the United States of America by Oxford University Press
198 Madison Avenue, New York, NY 10016, United States of America

British Library Cataloguing in Publication Data
Data available

Library of Congress Control Number: 2026936875

ISBN 9780198862253

The manufacturer's authorized representative in the EU for product safety is Oxford University Press España S.A. of Parque Empresarial San Fernando de Henares, Avenida de Castilla, 2 - 28830 Madrid (www.oup.es/en or product.safety@oup.com). OUP España S.A. also acts as importer into Spain of products made by the manufacturer.

Dedication

For James, and like-minded curious folk, who love nature and want to explore.

PREFACE

Welcome to the Oxford Biology Primers

There has never been a more exciting time to be a biologist. Not only do we understand more about the biological world than ever before, but we're using that understanding in ever more creative and valuable ways.

Our understanding of the way our genes work is being used to explore new ways to treat disease; our understanding of ecosystems is being used to explore more effective ways to protect the diversity of life on Earth; our understanding of plant science is being used to explore more sustainable ways to feed a growing human population.

The repeated use of the word 'explore' here is no accident. The study of biology is, at heart, an exploration. We have written the Oxford Biology Primers to encourage you to explore biology for yourself—to find out more about what scientists at the cutting edge of the subject are researching, and the biological problems they're trying to solve.

Throughout the series, we use a range of features to help you see topics from different perspectives.

Scientific approach panels help you understand a little more about 'how we know what we know'—that is, the research that has been carried out to reveal our current understanding of the science described in the text, and the methods and approaches scientists have used when carrying out that research.

Case studies explore how a particular concept is relevant to our everyday life, or provide an intimate picture of one aspect of the science described.

The bigger picture panels help you think about some of the issues and challenges associated with the topic under discussion—for example, ethical considerations, or wider impacts on society.

More than anything, however, we hope this series will reveal to you, its readers, that biology is awe-inspiring, both in its variety and its intricacy, and will drive you forward to explore the subject further for yourself.

CONTENTS

ACKNOWLEDGEMENTS

Writing this primer reminds me how exciting it can be to discover a new thing. If, like me, you find statistics daunting, remember that it is just a tool to help you explain to others what you've found; but know that it also empowers you to explore further and deeper than you ever thought possible. I am very fortunate to have spent my career working with people who use their days to think about life and how it all works, so my first thanks go to my biology-minded friends and colleagues who have kept my interests diverse and unlimited over the years. Particular thanks go to my colleagues in the D'Arcy Thompson Unit at the University of Dundee whose deep understanding of life sciences continues to inspire me, as well as the next generation of biologists. You can turn a day trip to the Dundee Botanical Gardens into a deep exploration of life, the universe, and everything!

This primer had its origins during the Covid pandemic and might never have seen the light of day were it not for Ann Fullick's outstanding guidance and unwavering encouragement. I am grateful to the experts that Ann asked to sense-check what I had written, particularly Gill Hickman whose comments helped to shape the final draft, and Chris Short who provided a reassuring check through the statistics. Most importantly of all, thanks to my wonderful wife, Fiona, and family Sophie, Ben, Luca, Bella, and Kasia for keeping me going with ever 'helpful' comments, genuine encouragement, and lots of fun times.

ABOUT THE AUTHOR

Professor Stephen Land PhD, CSci Teach, FRSB, FTPS

Stephen started out in zoology and marine biology laboratories in the UK (Aberdeen), Canada (UBC, Vancouver), and the USA (Marine Biological Laboratory, Woods Hole) that based their science in an idea called the August Krogh Principle: that many disease processes can be understood by studying animals that have evolved adaptations to survive in extreme conditions. For example, these labs studied how fish live in the low oxygen levels in the Amazon River, how Quechua people thrive in the thin air of the high Andes mountains, through to freeze tolerance and metabolic dormancy in freshwater frogs and turtles. With this background, and inspired by Joseph Barcroft's description of the low oxygen (hypoxic) environment of the womb as 'Mount Everest *in utero*', Stephen embarked on a research career to explain how hypoxia drives fetal lung development and sets the conditions for oxygen uptake from birth to adult life. This involved a search for the developmental clocks which tie the oxygen molecule to the branching growth of the airway and blood vessels that form the lung's respiratory tree. By exploring how the complex anatomy of the lung is formed, this work helped to shed light on lung defects in genetic diseases like tuberous sclerosis, Birt-Hogg-Dubé syndrome, and cystic fibrosis.

Stephen has a personal chair in Physiology and Biomedical Education at the University of Dundee, is a Chartered Science Teacher and Fellow of the Royal Society of Biology, and a Fellow of the Physiological Society. In his spare time, he likes to explore with his dog, look under things, and generally get very muddy.

ABOUT THE EDITOR

Ann Fullick

International biology author and education consultant, Fellow of the Royal Society of Biology.

Ann is a Cambridge Natural Sciences graduate who taught Biology for many years, ran a large Science department and examined at both GCSE and A level. She has written around 200 biology books for schools, both in the UK and around the world, and developed award-winning online biology resources. Ann has advised Government ministers in the UK and elsewhere on the development of biology curricula and worked with NGOs—most recently on the UNICEF Learning Passport and with NISSEM—supporting students globally where access to education isn't easy.

Ann is a member of the Biological Education Research Group and a Fellow of the Royal Society of Biology, where she has been actively involved in the Curriculum Committee developing a framework for biology education in the future. She has recently been awarded the Royal Society of Biology President's Medal for services to biology education.

1 WHY EXPERIMENTAL DESIGN AND STATISTICS MATTER

It is easy to imagine that it takes a genius or a grand research collaboration to make the discoveries that change the world, but most big breakthroughs in science history come from simple ideas.

Look out of the window—the Earth looks flat, doesn't it? That's how it would have appeared to the ancient Greeks, but as keen observers, they questioned the appearance of the world around them. They noticed that ships sailing towards the horizon disappeared hull-before-sail as if going over a curve. If the Earth was flat, surely the ship would stay in proportion as it sailed into the distance? They also noted that the height of constellations above the horizon increased with southward travel to Egypt and the trade routes of the River Nile. If the Earth was flat, wouldn't they stay in the same position?

Around 2240 years ago, Eratosthenes, a Greek librarian and mathematician in the ancient Egyptian city of Alexandria, conducted a simple experiment to test the idea that the Earth is a sphere. We didn't see that sphere with our own eyes (Figure 1.1) until 1972!

This primer aims to break the 'glass ceiling' view of scientific research by introducing a way of thinking that puts the power to discover new things into your hands. Here, you will discover how scientists use experimentation to make extraordinary insights into the world around us. You will explore statistics as a tool for explaining patterns within data and how this tool can be used for hypothesis testing, dispelling falsehoods, and promoting discovery.

Figure 1.1 It was over 2000 years after Eratosthenes' observations before the famous 'Blue Marble' image—taken by the crew of Apollo 17 on its way to the Moon in 1972—allowed everyone to see that he was right. The Earth really is a sphere!

How do we know what we know?

Science is always changing as our knowledge and understanding grows. Scientists aim to increase that knowledge and understanding, and observation, experimentation, and analysis are often key elements of successful work.

Returning to our Greek librarian, Eratosthenes—he reasoned that if the Earth's surface was curved, two points on a North-South line would have different angles to the parallel rays of the noon sun. If the Earth's surface was flat, these angles would be the same. He had heard that no shade was cast at noon on the summer solstice in the city of Syene (modern day Aswan). This city was 5000 stadia, or about 800 km, to the south of Alexandria along the River Nile. Eratosthenes had a stick of known length placed into the ground at Alexandria and another at Syene. He then arranged for the shadow length of each stick to be measured at noon on the summer solstice and compared them. From the observations, he measured an angle of 7.2° between the two cities and used this to calculate the Earth's circumference as 250,000 stadia or 40,000km in today's measurements. Eratosthenes showed that the Earth was (and is) indeed a sphere—a radical idea for the time!

Eratosthenes' measurement of the circumference of the Earth was simple but changed the course of human history. You can repeat his experiment using an apple, two cocktail sticks, and a torch (Figure 1.2 and Further reading).

The first image of the Earth from space which proved the Earth's surface was curved was taken in 1946 using a German V2 Rocket captured at the

Figure 1.2 Eratosthenes' experiment replicated using an apple, two matchsticks, and a torch. In Further Reading, you can follow a link from the National Aeronautics and Space Administration Goddard Space Flight Center to see how the method works. Try using it to calculate the circumference of the apple.

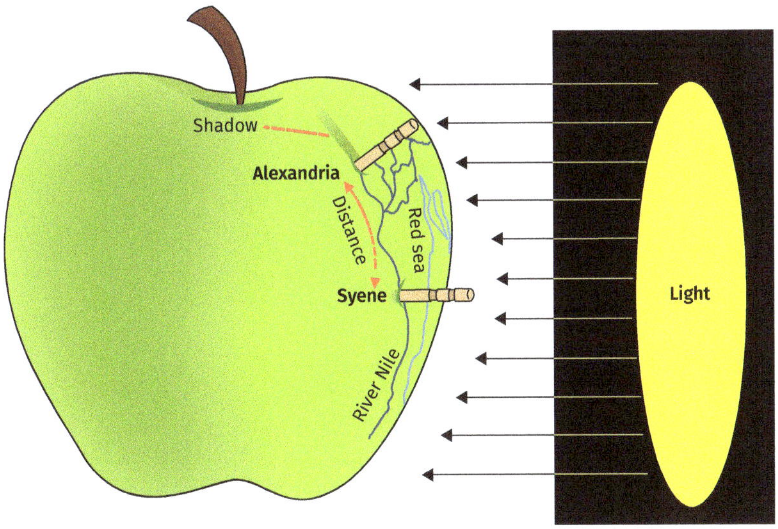

end of World War II. Today, space flight has enabled us to see the curvature of the Earth's crust with our own eyes and geo-positioning satellites have measured our planet's equatorial circumference as 40,075km. Eratosthenes' big idea, measured with sticks 2000 years before, was not only correct but his calculations were accurate to within 0.1%! The lesson is that observation and experimentation are powerful tools which can be used by anyone to challenge our views of the world. If Eratosthenes could measure the circumference of the Earth using sticks, imagine what you could do with the tools we have in the modern age!

The road to the Scientific Method: from Aristotle to Popper

> 'Reports that say that something hasn't happened are always interesting to me, because as we know, there are known knowns; there are things we know we know. We also know there are known unknowns; that is to say we know there are some things we do not know. But there are also unknown unknowns—the ones we don't know we don't know.'
>
> US Defence Secretary, Donald Rumsfeld in 2002

The quote above (see Figure 1.3) presents a sceptical point of view because it shows that all points of view, known and unknown, should be considered when addressing a problem. A sceptical approach is as relevant to scientific exploration and planning as it is to the world of spies and intelligence gathering.

Figure 1.3 Donald Rumsfeld (US Defense Secretary, Bush Administration) put forward a sceptical argument to journalists to make his case for the Iraq War in 2002. Source: File:US Navy 030825-F-2828D-352 Secretary of Defense Donald H. Rumsfeld responds to a reporter's question during a press conference at Lackland Air Force Base, San Antonio.jpg – Wikimedia Commons

This provides a powerful approach to questioning the way we understand things because it allows ideas to be challenged by anyone, anywhere, at any time. The scientific method (Figure 1.4) provides a framework for testing ideas where our ability to question more deeply becomes stronger as evidence grows. It minimizes the influence of myth, legend and received wisdom—such as the Norse Sagas, Greek Mythologies, or mainstream religions—in knowledge building, by placing evidence gained by observation and experimentation at the heart of the discovery process.

Figure 1.4 Eratosthenes used a form of systematic thinking which has become known as 'The Scientific Method'.

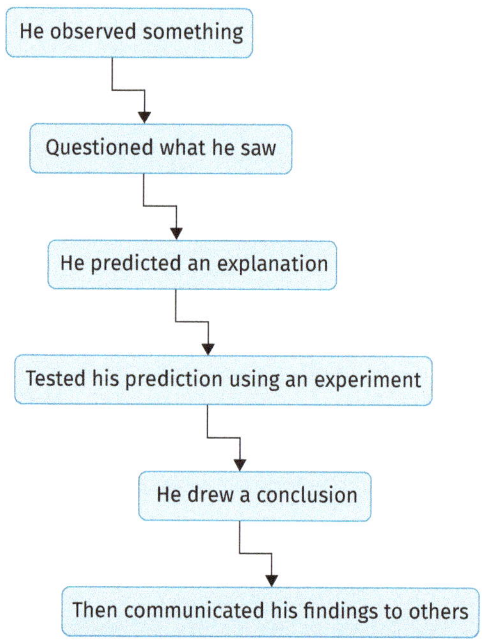

This type of systematic thinking originates from a kind of philosophy known as **scepticism**, where knowledge is developed by weighing the evidence for and against an idea before conclusions are drawn. Scientific scepticism uses the scientific method to put ideas through a process of evidence-based investigation. Curiosity about the world around is what makes us human and so this approach has developed in different ways and times in cultures around the world. Figure 1.5 highlights eight European thinkers who wrote down their views in a way which helps us see how the scientific method has been refined over 2400 years and, so, has changed the way we think about the world around us.

Figure 1.5 The scientific method has evolved across diverse human cultures over 2000 years and more. This timeline introduces eight key European thinkers whose beliefs have shaped our approach to modern scientific investigation and which drive its refinement into the future.

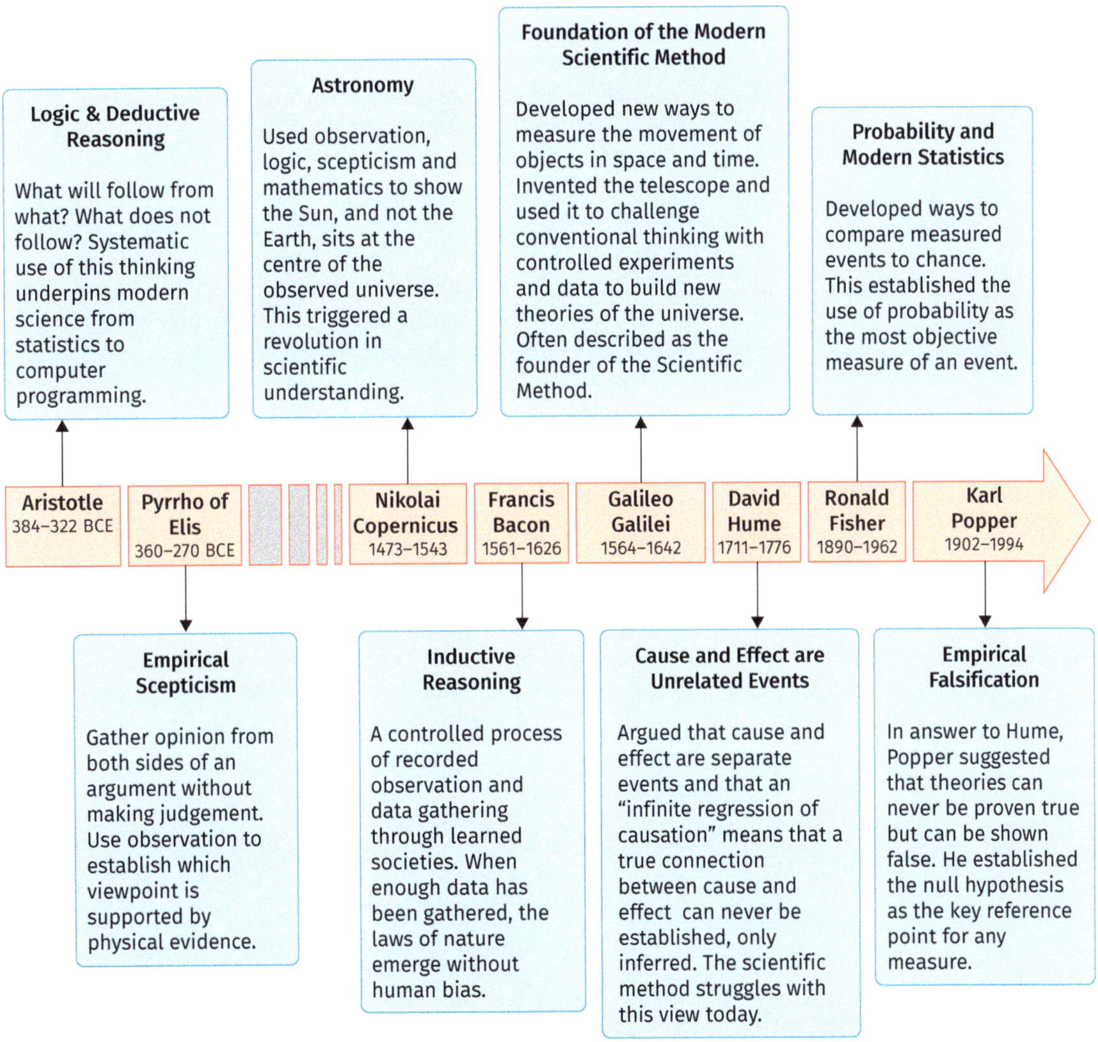

Female pioneers of scientific thinking

All the people highlighted in Figure 1.5 are men, so where are the women? Women often feature less prominently in the history of the scientific method, because opportunities for education and the advancement of scientific views traditionally favoured men. However, evidence is emerging to show that women have made essential contributions to scientific thinking for at least 4000 years. In ancient times, many were prominent physicians or chemists who worked to understand the nature of medicines and perfumes. In the 18th century, Marie-Anne Paulze Lavoisier was instrumental in the standardization of the scientific method and in bringing together the different disciplines which make up the field of chemistry, moving it away from the "hocus-pocus" beliefs of the alchemists. She worked with her husband before his death and carried on her scientific endeavours and writing after he was executed in the French revolution. In the 19th and 20th centuries female scientists made vital advances in scientific thinking from developing the origins of medical statistics and the importance of good hygiene practices in nursing (Florence Nightingale and Mary Seacole), understanding of atomic structure and radiation (Marie Curie), to the origins of life on Earth (Lynn Margulis). You can discover more about the historical role of women in science from the further reading at the end of this chapter.

The modern scientific method

In the modern world, the scientific method works rather like a whirlpool where ideas are tested by predicting outcomes based on the evidence available. The results inform further experiments until enough evidence is gathered to build theories which underpin the laws we use to predict the behaviour of the world around us.

The following terms are commonly used in following the scientific method:

Term	Definition
Observation	A process of information gathering using the primary human senses of sight, sound, touch, taste, or smell.
Hypothesis	The prediction of an outcome from an experimental manipulation.
Null hypothesis	The prediction of no effect or outcome from an experimental manipulation.
Independent variable	A condition which is manipulated by the experimenter.
Dependent variable	An effect which is predicted by the hypothesis to respond to manipulation of the independent variable by the experimenter.
Control variable	An effect which does not respond to manipulation of the independent variable by the experimenter.

Data	Information which is gathered to record the effect of manipulating the independent variable upon the dependent and control variables.
Conclusion	A decision upon whether to accept or reject the hypothesis which is made by examining the difference in effect data between the dependent and control variables.
Replication	Repeat of an experiment to determine if the observed outcome can be obtained more frequently than might be predicted by chance alone.
Theory	An explanation of an observed event which is based on evidence gathered from systematic hypothesis testing, but which is, itself, open to challenge from refinements made to the scientific method.
Law	A principle which is based upon one or more theories which can be used to predict, but not explain, the behaviour of an interaction between independent and dependent variables.

This list gives some of the terms that scientists use to communicate the process of evidence gathering by experimentation, which in turn is used to construct the theories and laws on which our knowledge is based. Take Mendel's Laws of Heredity, for example. Gregor Mendel (1822–1884) was an Augustinian friar living in the town of Brno in present-day Czechia. His life-long fascination with the characteristics of peas laid the foundation for modern understanding of genetic inheritance! Between 1856 and 1863 he experimented by hybridizing plants with different characteristic traits (independent variables) and recorded data to report the effect of these experiments upon seed shape, seed colour, flower colour, flower position, and plant height (dependent variables). He replicated his experiments at different time intervals to establish whether his observations were more than mere chance. He used nearly 28,000 plants to develop his theories to explain how traits are passed from one generation to the next—a substantial data set! From these observations came Mendel's Laws of Heredity with which you may well be familiar:

Law of Genetic Segregation—Inherited traits are controlled by pairs of alleles. A gene is a section of DNA that affects a trait, and an allele is a different version of that gene. These allele pairs separate into sex cells so each cell gets only one. At fertilization, offspring receive one allele from each parent to form a new pair.

Law of Independent Assortment—Alleles for different traits are not dependent upon the inheritance of one another.

Law of Genetic Dominance—Dominant alleles are always expressed in preference to recessive alleles.

Mendel's Laws make predictions that can be used to explore unknown questions. They do not explain how a process works but, by predicting possible outcomes, they enable scientists to determine if an observation fits a known theory. In Mendel's case, he had no way of knowing whether the process of heredity worked the same way for peas as for other plants and animals and it was to be 100 years before the discovery of DNA, chromosomes, genes, and different alleles provided the molecular explanation for his theories. Generations of scientists after Mendel established that his laws can be used to predict the outcome of sexual transfer of genetic material from one generation to another for almost all living things. Where exceptions occur, we discover new ways in which the process of heredity operates which expands our understanding of genetics.

You may be surprised to learn that Mendel's experiments were mostly ignored by other scientists at the time, largely because he developed his laws without knowledge of the biological mechanism that underpinned them. It was easy for the influential scientists of the day to be sceptical about the work of a relatively unknown monk. Sixteen years after his death, however, his work was replicated and slowly gained acceptance over the following half-century. Today, through replication in different areas of biology, Mendel's Laws of Heredity are accepted as fundamental to the process of genetic inheritance.

Scientific approach 1.1

Using the scientific method to find out if bees can see colour

How do we know if honeybees can see colour? We assume that they might because, after all, the flowers that bees feed from are always coloured in some way (Figure A). Except for a few species of woodland flowers which become transparent in the rain, there is no such thing as a truly colourless flower.

So, we have our **observation**: flower pigment seems to be important for attracting bees because they feed exclusively from coloured flowers. Our observation doesn't provide any scientific explanation because it does not tell us if bees are attracted by colour, tonal shade, or part of the light spectrum that we cannot see. It also ignores the possibility that flower scent, shape, or position may play a role. So how can we test if bees can respond to the colour of a flower?

Following Popper's idea, we need a null hypothesis which can be tested to show which parts of this observation are false. A simple **null hypothesis** would be:

'Honeybee feeding behaviour is not influenced by colour.'

To test this, we will use an experiment devised by the Austrian behavioural biologist Karl von Frisch in 1914 to show that bees can see colour (von Frisch won a Nobel Prize in 1973 with two other famous behavioural biologists,

Figure A: Honeybees are a familiar sight on flowers—but how important is the colour of the flower to attracting the bee?

Konrad Lorenz and Nikolaas Tinbergen for their discoveries of animal and human behaviour patterns). This deceptively simple experiment only requires five squares of glass (e.g. 10 × 10 cm) with a selection of coloured cards cut to the same size, a white cotton tablecloth, a 1:10 dilution of honey in water, a fine-tipped paint brush, and quick drying plastic paint (any bright colour).

In this experiment, the **independent variable** is colour, the **dependent variable** is honeybee feeding behaviour, and the **control variable**, as we will see, is grey tone.

The five sheets of card are laid out on the white cotton cloth (plastic diffracts the light and may disrupt the experiment) and each is covered with a separate sheet of glass. Each card colour reflects blocks of colour from the visible spectrum such as red, yellow, green, blue, and purple.

A small amount of honey solution diluted 1:10 with water is placed on the glass covering the blue card and a bee is gently transferred from a flower onto the blue plate next to the diluted honey solution. When the bee begins to feed on the honey, it is gently tagged on the back of the head with a very small dot of coloured plastic paint. After feeding, the bee flies back to the hive and informs others of its find using the 'waggle dance' (Figure B) that von Frisch showed to be a means of communicating the location and distance of a food source to other hive members.

The tagged bee returns to the blue glass plate with other members of its hive, which are tagged in the same way. When they leave, the honey is removed and a clean sheet of glass substituted. The tagged bees return exclusively to the blue sheet of glass and will ignore all other shades of colour offered elsewhere on the table.

To dissect the bee's visible spectrum, the experiment is repeated using other colours. The response appears strongest towards the blue end of the spectrum and weakens towards the red end.

Figure B: Honeybees communicate information about food sources using a 'waggle dance'. Karl von Frisch showed that the angle of the dance and frequency of the waggle indicated direction and distance to a food source to other members of the hive. His experiment with coloured plates showed that honeybees associate particular colours with food sources.

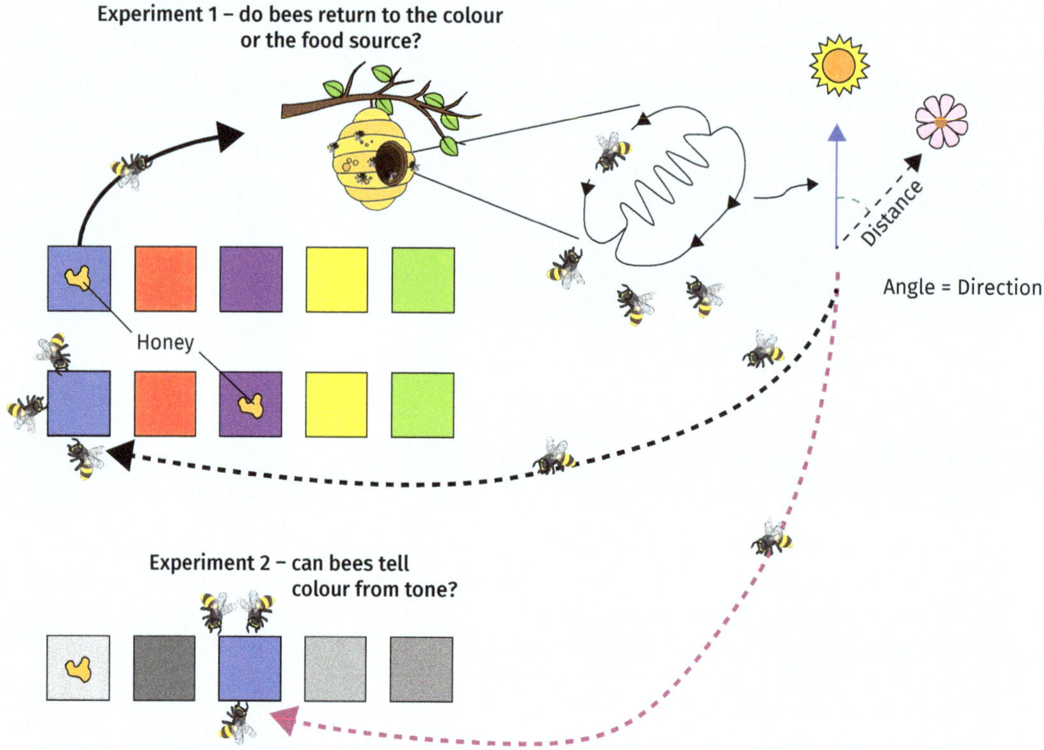

So far, this experiment does not, in fact, prove that bees see colour, only that they alter their feeding behaviour in response to shades of tone. To prove that bees respond only to the colour, the experiment is repeated with one coloured card surrounded by others with different shades of grey. The bee will return to the coloured card without confusing it for grey colours of similar tone. The one exception is the colour red which bees cannot distinguish and will confuse with grey tones.

The philosophers in Figure 1.5 provide us with some interesting thoughts about what the results might show.

Following Popper's philosophy of empirical falsification, we see that bee feeding behaviour does indeed respond to colour. We can conclude that the experiments show the null hypothesis to be false and that honeybee feeding behaviour can be demonstrated to be sensitive to colours other than red.

Hume's sceptical view of cause and effect means that this experiment does not say anything about the relationship between flower colour and honeybee feeding; we cannot say that flower colour causes honeybee feeding, only that flowers are coloured, bees can see colour, and their feeding behaviour can be influenced by colour.

Over time, von Frisch performed many complex behavioural experiments to show how bees communicate food source locations to hive members, how they navigate using the changing position of the Sun over the course of a day, their perception of polarized light beyond the visual spectrum of the human eye, and their use of the magnetic field of the Earth to orientate the position of honeycombs.

Pause for thought

- Flowering plants use scent as well as colour to attract pollinators. How might the properties of scent help plants to communicate with pollinators in different ways to colour?
- Suggest how you might develop von Frish's approach to test your ideas.

How do scientists think?

Of course, scientists are all individuals and think in very different ways, but there are common strands to the ways they tackle problems. Let's take von Frisch's honeybee experiment in Scientific Approach 1.1 and use it to show how scientists use different reasoning strategies to develop hypotheses, theories, and laws.

The research process begins with a process known as **abductive reasoning**, which is used to develop a basic pattern of evidence, rather like assembling scattered pieces of a jigsaw to gain a sense of what the overall picture might look like. In our honeybee example, abductive reasoning would be used to develop a set of observations to explore the relationship between bees and flowers:

***Question*:** Why do honeybees visit gardens?

***Observation 1*:** Gardens with flowering plants contain more bees than those with no flowering plants.

***Observation 2*:** Bees land on coloured flowers within the garden.

***Assumption*:** Flowers seem to have something to do with attracting bees into gardens. We might imagine that this could have something to do with flower scent, colour, height, or shape.

Scientists often use abductive reasoning in pilot studies to generate a basic line of evidence that can be used to develop a research strategy.

The next stage involves **inductive reasoning** to develop a hypothesis that can be used to test ideas, generate data, and form a set of conclusions. In going from 'specific observation' to 'general conclusion', inductive reasoning is dependent upon the reliability of the starting evidence and allows for conclusions to be false, even if the starting **premise** is believed to be true. Inductive reasoning was championed by Francis Bacon (Figure 1.6) and begins with an observation supported by empirical evidence (data) from which a conclusion is drawn. In the honeybee experiment inductive reasoning might run as:

> ***Observed premise 1:*** 7 out of 10 bees which have received honey from a blue coloured plate will return to it on a later occasion even if there is no honey available.
>
> ***Observed premise 2:*** 9 out of 10 bees will return to the blue plate in preference to any other colour or grey shade if they have received honey from it on a previous occasion.
>
> ***Conclusion:*** feeding behaviour of most bees is influenced by the colour blue.
>
> ***Replication:*** Evidence that colour is important to bee behaviour and may explain the relationship between flower colour, bee feeding habits, and the reproduction of both bees and flowering plants is universal. These premises will be found true at any future time and place.

Note how the conclusions do not take account of other factors such as flower scent, shape, height, or weather, which could just as easily account for bee feeding behaviour. Although the evidence is reliable, the conclusion stated here could be wrong and would require strengthening by further experiments. Inductive reasoning is at the heart of most scientific experiments and is used to develop the theories and laws.

The final stage involves **deductive reasoning** which begins with a general principle or '**inference**' from which a set of **logical statements** can be made towards a predicted conclusion. In the context of science, a 'premise' is the

Figure 1.6 The scientific method employs a hierarchy of reasoned thinking: abductive reasoning to formulate questions, inductive reasoning to test them, and deductive reasoning enabling understanding gained in one area of science to be applied to new unknowns.

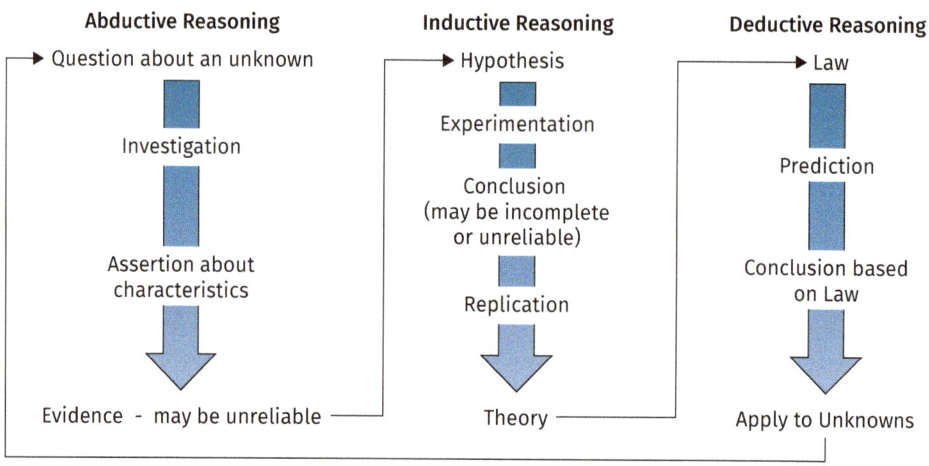

same as a law. The reasoning process flows from 'general law' to 'specific conclusion' and so the strength of the deductive process depends upon the reliability of the law. Karl von Frisch's experiments on honeybee migration established that the waggle dance of honeybees indicates the direction and distance to a food source. We can refer to this as a law because it can be used to predict the feeding behaviour of bees, and so in our honeybee example, deductive reasoning flows as:

***Law (inference)*:** Hive members in contact with a bee performing the waggle dance will always travel to the location of the food source that was last visited by the dancing bee.

***Reasoned statement 1*:** Food sources are spread out and have different characteristics, so bees require more information than simply distance and location to find the same food source as the dancing bee.

***Reasoned statement 2*:** Bees recognize flower colour.

Conclusion: The waggle dance must include information about flower colour.

Deductive reasoning is powerful when the premise is based on a theory or law that is shown through experimentation and replication to be reliable.

Scientists use abductive, inductive, and deductive reasoning throughout the different stages of the scientific method (Figure 1.6). Abductive reasoning provides a starting place, inductive reasoning is used to form testable hypotheses and reliable theories, whereas deductive reasoning allows laws to be applied to wider questions about the world around us. As theories and laws are challenged by experimentation, our perception of what is 'true' changes with time.

Case study 1.1

The 1835 Nuremberg salt test

Let's look at an example of how the scientific method has been used to dispel myth. From the time of the Ancient Greeks to the present day, health spas have promoted the idea of '*salus per aquam*', health from water. In the past, when there were far fewer effective medical treatments such as antibiotics, people were often desperate for cures and some of the claims about the healing powers of 'water' were far from scientifically valid. In the mid-19th century, in the Spa town of Nuremberg in Germany, there was an episode which demonstrated very publicly the power of the scientific method to expose false claims about health treatments. In this case, the evidence was against the practice of **homeopathy**, where people were treated with heavily diluted potions of herbal cocktails, so dilute that they were in fact just water!

Friedrich Wilhelm von Hoven was a leading doctor and public health official in Nuremberg. He was concerned about the grip that homeopathy had on treatments for ailments. Physician homeopaths often treated the aristocracy, who could afford to pay and who did not want to appear foolish or admit failure. Von Hoven was concerned that this obsession with homeopathy would stop the progress of medical discovery based on evidence and observation, a concept now known as '**evidence-based medicine**'. Fearing he might lose his position by criticizing powerful homeopathic physicians with their ties to the rich aristocracy, von Hoven published his sceptical views under a false name. As he suspected, his ideas drew the anger of the leading physician homeopath in Nuremberg, Johann Jacob Reuter. Reuter challenged von Hoven to try the effects of a solution of salt that had been diluted one part in 10, 30 times over (Figure A). The idea was to create a tincture of pure water that retained a 'memory' of the substance that had originally been dissolved within it.

All homeopathic remedies are created by putting an active ingredient, in this case salt in water, through a series of sequential 10-fold dilutions (shown as 'C'). The 1835 Nuremberg salt test used one grain of salt diluted in 100 drops of distilled snow water which was then diluted 29 times to create a C30 solution. This dilution procedure would have resulted in the complete clearance of salt molecules from the solution after C23. The belief is that the homeopathic benefit of such extreme dilution results from the suggestion that water retains a 'memory' of the substance that had been diluted within it.

The claim made by Reuter was that the tincture produced by his salt dilution would improve the health of any person who was sick, a perfectly reasonable hypothesis which lends itself to investigation using the scientific method.

Having revealed his identity as the source of the sceptical attack, von Hoven accepted the challenge on the condition that the trial involved as many participants as possible and that the identity of the treatments was hidden to all the participants including the experimenters. One hundred and twenty townspeople gathered in a local tavern where the treatments were prepared in front of them.

Figure A: The dilution protocol for the 1835 Nuremberg salt test - supporters claimed that even after the original salt had been diluted to the point where no molecules remained, the water retained a beneficial "memory" of the substance that had once been dissolved in it.

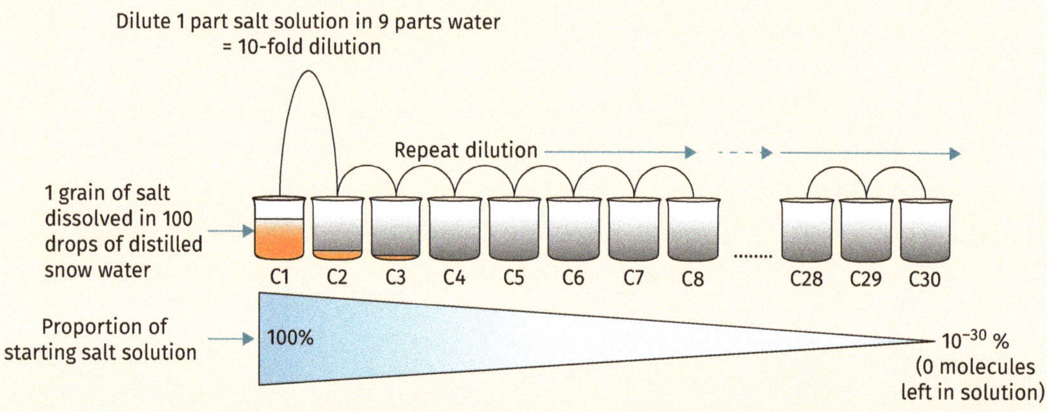

- First, 100 empty vials were numbered then shuffled into random order.
- One half were filled with distilled snow water (the control variable, or 'placebo').
- The other half were filled with C30 salt diluted in distilled snow water (the dependent variable).
- Two independent pharmacists were asked to prepare the treatments. To ensure that the risk of contamination was minimal, they were prevented from working with other medicines for two days before the trial, then had to take a bath (unusual in those days) before preparing the solutions using brand-new weighing scales.
- An independent commission then distributed half of the randomized vials among 50 volunteering townsfolk and made a secret list of who had received the treatments.
- Three weeks later, each volunteer was asked to report any symptoms or physical experiences that had occurred after drinking the solution.

And the results? Eight of the 50 participants reported some effect on their health but of these, three had received the water placebo! Ninety per cent of those who took part experienced no effect regardless of the treatment they had received.

This study was important because it gave equal chance to all possible outcomes by minimizing sources of bias: the vials were randomized, and the trial was double-blind so that neither the participants nor the experimenters knew who had received either treatment. Importantly, the trial was public, and any irregularities were recorded (for example, failure to report outcomes or pre-knowledge of vial contents despite the randomization protocol). The first randomized double-blind trial, based around the scientific method, had been born.

? Pause for thought

- What elements of this ground-breaking investigation show good scientific practice—and are there limitations by modern standards?
- Many people still practise homeopathy—and even more people use it—in spite of evidence such as the case study described here. Discuss some of the reasons why homeopathy remains so widely used.

Why probability rather than certainty?

This might seem an odd thing for a scientist to say—but science cannot establish truth. Why? Because the scientific method works by constantly challenging theories and laws to expose exceptions to the rule. From these exceptions, we build new lines of enquiry and deepen understanding. Theories are based on our understanding of things based on evidence *at the present time* but we cannot rule out the possibility that some future discovery might change what we currently think is 'true'. The 20th-century Philosopher, Karl Popper, was aware of this problem and proposed a theory of Empirical Falsification which states that we might never reliably show things to be true, but things can, with high certainty, be shown to be false. To illustrate this, he wrote:

> 'no matter how many instances of white swans we may have observed, this does not justify the conclusion that all swans are white.'

It doesn't matter how many white swans you see; it doesn't prove that all swans are white. But you only have to see one black swan to know for certain that all swans are NOT white! Of course, today we know *Cygnus atratus* is a species of black swan native to Australia, showing us that the white swan hypothesis is empirically false (Figure 1.7).

Popper's empirical falsification theory also helps us to understand the unpredictable nature of discovery. Although Europeans did not travel to Australia to specifically disprove the white swan hypothesis (that would be ridiculous!), its falsification arose from an extraordinary sequence of events, each involving their own application of the scientific method in some form:

- The political and economic drive to explore.
- The development of seaworthy ocean-going vessels.
- An understanding of the movement of planets and the mathematics to use this for navigation.
- The development of universally reliable navigational tools (sextant, chronometer).
- The charting and recording of navigable routes between continents.
- An internationally agreed order of cataloguing the relatedness of living things so that unknown species might be compared to those which are known.
- A method for communicating discoveries to the rest of the world.

Figure 1.7 A. The well-known European mute swan, *Cygnus olor*, typically has white plumage. B. The Australian black swan, *Cygnus atratus*, has black plumage but was not known to European explorers until they found the continent of Australia. Credit: Anthony Short

A B

Think for a moment how the world has been influenced by these events. The world of science is full of unexpected discoveries that have sprung from scientific exploration (Figure 1.8).

For this reason, scientists do not think in terms of 'true' or 'false' but instead consider the likelihood that an event might occur more times than by chance alone. Popper showed that although it is not possible to establish truth, it is quite possible to use the scientific method to demonstrate that things are false more often than one might expect by chance. This argument placed the null hypothesis—that an independent variable has no effect upon a dependent variable—at the centre of scientific investigation. Using probability to determine if an effect occurs more than might be expected by chance to dispel the null hypothesis advances understanding through the scientific method.

What makes science reliable?

So far, we have learned three important things:

1. The scientific method is a template for systematic enquiry which helps us to test ideas through the process of hypothesis setting, experimentation, and the use of observations to challenge laws.
2. Philosophers argue that cause and effect are independent things so that ultimate truth can never be established. Facts can never be proven, only shown to be false.
3. Bias is part of human nature. People tend to search for explanations that they believe to be true even if the scientific evidence shows otherwise.

This is pretty confusing! How can science reliably tell us anything in the face of these opposing statements?

Figure 1.8 The microwave oven arose from a chance discovery by Percy Spencer as he experimented with microwave radiation to develop radar in 1945. During one of his experiments, he discovered that the waves had melted a chocolate bar in his pocket. Microwaves were briefly explored as a way of warming hypothermic human tissue in a process known as diathermy. Nowadays, they're mostly used to warm food.

Objectivity in science

Objectivity is difficult to define. Noah Dongen and Michal Sikorski are research philosophers who asked other scientists if they understood the meaning of objectivity in their work. The scientists identified things which are features of objectivity, such as accuracy of data recording, reproducibility of data, and lowering risk of reporting false positives, but few could agree on a working definition of objectivity. In their 2021 article called 'Objectivity for the Research Worker' Dongen and Sikorski proposed a working definition with major implications for how science is done in the future. They said:

> 'A scientific practice becomes more objective when it becomes demonstrably more resilient to actions and decisions that have the potential to influence its outcome; concretely when:
>
> a) The study design and data collection become demonstrably more resilient to the scientists' influence on the data
>
> b) And the data processing and analysis become demonstrably more resilient to ad hoc decision making and selective reporting of positive results.'
>
> Dongen & Sikorski
> *European Journal for Philosophy of Science* (2021) 11:93

In other words, people are the problem when it comes to questions of reliability in science! Scientists build objective practice into an experiment by testing if it can be replicated more often than predicted by chance, by any researcher working anywhere and at any time. This is not to say that scientists are constantly reproducing each other's work—this would slow the progress of discovery to a crawl—but by publishing their research they allow other scientists to scrutinize the reliability of their discoveries (Figure 1.9). Here are some of the built-in protections used to maintain objectivity in scientific research.

- To limit risk of human bias, treatments are ideally randomized and conducted without the person conducting the experiment knowing which treatment has been given where.
- Data analysis should be performed by a person with no vested interest in the outcome of the study.
- Researchers should openly declare all conflicts of interest, such as any financial gain arising from the study.
- All scientific publications provide the information needed to replicate the work by including a methods section that describes the design of the study and the techniques used to measure outcomes.
- Statistical analysis is used to describe patterns within the data and establish if the observations have occurred more often than would be expected by chance alone.

Figure 1.9 Scientific publications can vary in structure but are usually organized into seven main sections, shown here in red text, with quality assurance information shown here in blue text

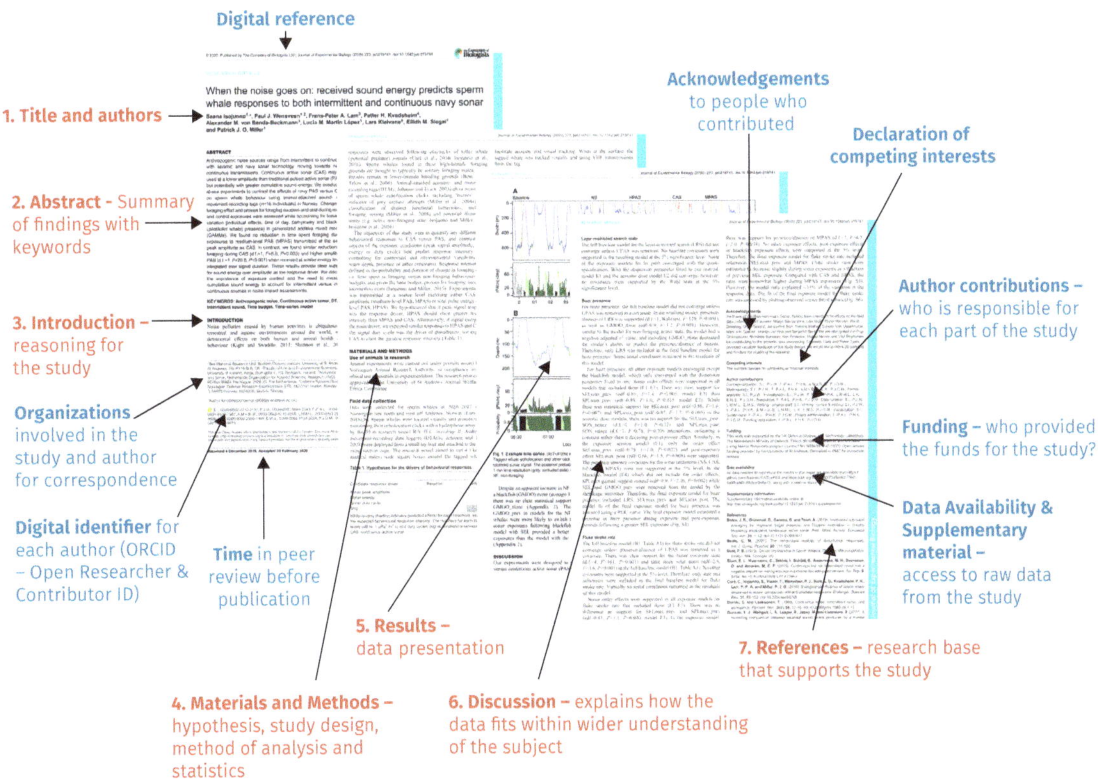

- When scientists submit their work for publication in a journal, it first goes through a process of scrutiny, or **peer review**, by other scientists who have knowledge of the field. A cycle of 'research-replication-peer review-publication-new research' builds up the objective reliability of an observation.

With several thousand articles appearing in life science journals every year, how do we know if a new publication paper is reliable? We know from Popper's swan example that it is not always possible to predict the effect that a new discovery might have for the direction of a field of research and so scientists use indirect indicators of influence about the authors and the reputation of the journal to determine how important a new publication might be. The impact that an individual scientist has had upon a field is measured through a value known as the *h*-index. This measures the number of times their work has been referenced (cited) by other scientists against the number of articles the scientist has published over their career (productivity). A high *h*-index suggests high productivity with a high rate of citation by other scientists; this person likely has a continuous and influential

Bigger picture 1.1

Tragic roots: statistics and eugenics

At the turn of the 20th century when genetics and the power of data-driven science were not well understood, a group of three statisticians called Francis Galton (the second cousin of Charles Darwin), Karl Pearson, and Ronald Fisher advocated for statistics to be used as a systematic method for removing undesirable inherited traits from the human population. This idea of racial purification came to be known as **eugenics**. It was adopted by the Nazis in Germany in the mid-20th century (Figure A) to justify the persecution and mass murder of Jewish people, minority groups, and people with

Figure A: A poster from the Third Eugenics Congress held in New York City in 1932. Eugenicists believed that statistics was an essential tool to study how different sciences could be used to create an idealised human race.

THIRD INTERNATIONAL EUGENICS CONGRESS, NEW YORK CITY, AUGUST 21–23, 1932
Introductory Wall Panel "The Relation of Eugenics to Other Sciences," based on a paper by Dr. Harry H. Laughlin, Cold Spring Harbor, Long Island, New York

disabilities. Members of the Nazi regime responsible for these atrocities were tried and convicted during the Nuremberg Trials at the end of World War II and the practice of eugenics was thoroughly discredited when the biased, non-objective scientific practices which had supported it were exposed. Despite their historical misuse, the statistical methods founded by Galton, Pearson, and Fisher remain in use to the present day and have benefitted understanding of nature and human society in lots of ways. You can find out more about the difficult history of statistics and its emergence as a powerful force for good in the further reading section.

Pause for thought

- The process of scientific enquiry means that our understanding of the world about us is constantly evolving but is also subject to the social views of the age. Think of one great leap in scientific understanding that has occurred in the last 150 years (splitting the atom, or discovery of DNA structure, for example). Draw a timeline from its discovery to the present and mark dates along its length where positive or negative outcomes have occurred.
- Do you think the scientists who made these discoveries could foresee how they might be used in the future? Do you think humans have much control over how scientific discoveries are used for good or bad from one generation to the next?

Case study 1.2

The measles–mumps–rubella vaccination scandal: how non-objective, biased science affects us

When thinking about the consequences of non-objective science, it's best first to ask, 'What is science for?' You might think, for example, that the purpose of science is to improve the lives of all people on the planet. Perhaps it is to safeguard the natural world and limit the effect that people have on other living things. Maybe it is about defining the laws that govern the universe so that we can understand our place within it. Whatever your answer, you will probably find that science impacts your life in many ways, as it does for billions of people around the world. So, whatever you believe science is for, it matters on a scale that can shape the fortunes of all living things on the planet.

Non-objective science

For centuries in the UK measles and mumps made children unwell—but had the potential to cause serious complications, brain damage, and even death. Rubella is relatively mild—but if a pregnant woman catches the disease, her foetus may be severely damaged or even born dead. In the UK in 1988 the combined measles–mumps–rubella (MMR) vaccine was introduced which protected children—and their mothers—against these diseases.

In 1998 an article was published in *The Lancet*, a leading medical journal with a high impact factor, which hypothesized that the MMR vaccine could cause autism in some children. The paper was based upon a study of 12 families whose children had pre-existing health conditions. Eight of these families reported the appearance of bowel conditions and autistic behaviour which seemed to develop days to months after the triple MMR vaccine had been given. This led the authors to recommend that the combined MMR vaccine should be given as separate shots, a step which would have greatly increased the number of times a doctor would need to see individual patients to administer each single dose and follow-up boosters.

This suggestion seems harmless enough, but the public reaction caught the scientific community by surprise. Uptake of any kind of vaccine, triple or single dose, fell below the level required to maintain community or 'herd' immunity and the incidence of measles and mumps, both potentially lethal diseases, began to rise in the UK population (Figure A). Mixed messages from the medical community, politicians, and media about the apparent safety of the triple MMR vaccine fuelled the debate. The effects rippled around the world, much like a game of 'Telephone' where facts become separated from truth as information passes from one community to another.

Financial incentives

In 2004 it emerged that the lead author of the original *Lancet* paper, Andrew Wakefield, had received a cash incentive from a legal aid group to find evidence that vaccines are more harmful than manufacturers claimed. An enquiry established that blood samples had been taken from children without consent and that some of the children in his tiny sample had not received the MMR vaccine at all. It was concluded that some of the data had been falsified to give the impression that a link between MMR and autism existed. Worse than non-objective—this was fake science, and very damaging. The work was retracted by *The Lancet* on the grounds that it was fraudulent, and that a major conflict of financial interest had not been declared.

Despite the scientific and medical community taking steps to highlight the major flaws in Wakefield's study, and Wakefield himself being disgraced and struck off the medical register as not fit to practise medicine, public confidence in the MMR vaccine was damaged and made worse by biased and sometimes almost hysterical media coverage. Several large, objective follow-up studies showing that the MMR vaccine carried very low risk of complications failed to reassure the public. After all, stories about safe vaccines do not generate the media interest that possible problems do! When faced with medical advice to vaccinate their children at two years of age, many parents chose not to do so—forgetting the severity of the diseases the MMR protects against (Figure A).

Figure A: As herd immunity fell in the wake of the MMR scandal, so the number of people infected with measles and mumps increased. The top graph shows the number of confirmed cases of measles, mumps, and rubella in England and Wales since publication of the article by Wakefield (indicated by a blue bar) which suggested the triple MMR vaccine might be linked to autism. The lower graph shows how the MMR immunization rate fell from 1998 to below the threshold for herd immunity. Source: Generated graphs using data from the Office of National Statistics https://www.gov.uk/government/publications/measles-confirmed-cases/confirmed-cases-of-measles-mumps-and-rubella-in-england-and-wales-2012-to-2013

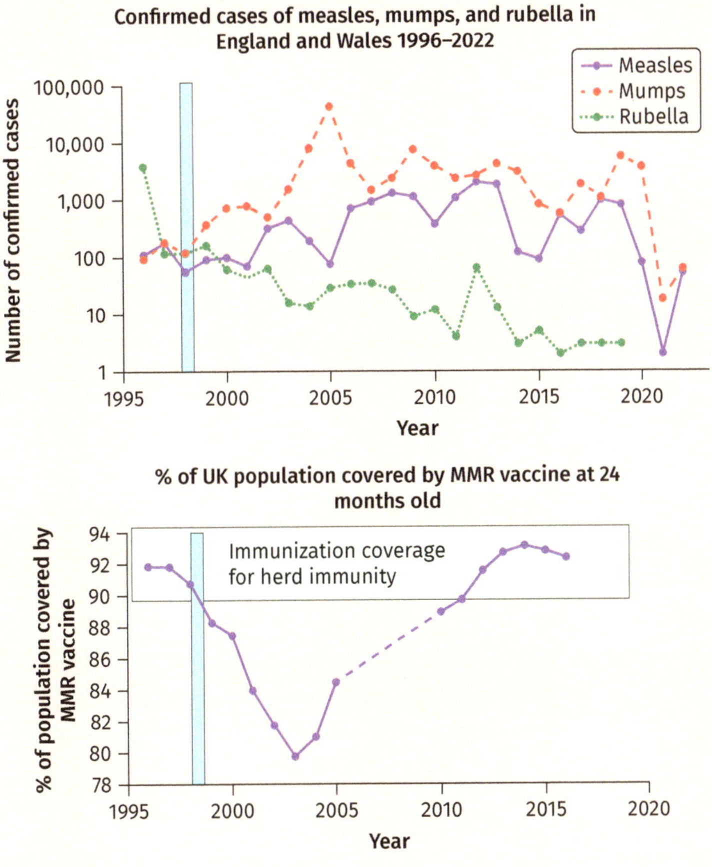

What are the lessons? Scientific discovery does not occur in a vacuum but has real consequences for individuals and communities. When conducted with integrity, science provides people with the ability to make decisions based on the best available evidence, but it does not claim to report certainty or establish beliefs. Non-objective science leaves the public unable to make informed choices because it undermines trust in the process of hypothesis testing, evidence gathering—and in the motivation of the scientists. With nowhere else to turn, some decide to make choices based on information from unreliable sources—made worse in this age of social media and fake news—and so trust in science becomes difficult to regain.

Pause for thought

- Social media is often accused of spreading inaccurate ideas and damaging conspiracy theories, but do you think it has a role to play in communicating objective science? Do you have examples of either good or bad practice involving social media and science?
- What sort of built-in objectivity measures would you expect to see on an internet site or article which would give you confidence that the information within was reliable?

impact on the field. A low h-index may indicate a scientist who is in the early stages of their career or who has been productive, but their work has not attracted attention for one reason or another. Journal reputation is measured by impact factor as the number of times all articles published in that journal have been cited in a year divided by the total number of articles published over the same time frame. Journals with a high impact factor indicate that the articles which have gone through that journal's peer review process are likely to have greater influence upon a field compared to journals with a low impact factor.

In this introduction we have looked at the thinking and processes which lie behind the scientific method. The next step is to explore how statistics are used as a tool to describe data and explore patterns which enable us to ask ever more penetrating questions. In a world without certainty, it is very important to use measures of probability to establish which lines of questioning will take us closer to the truth. To explore this, in the rest of this primer we will examine the basis of statistical thinking, how statistical tests are performed, the importance of designing experimental approaches that minimize bias, how to work with risk, and quantitative approaches to understanding function from measuring size and shape.

Chapter summary

- Some of the greatest ideas in science have come from well thought-out simple observations.
- The scientific method has developed as a process of experimental enquiry over more than 2000 years of philosophical thinking.
- Ideas can be falsified but never proven as certainties, so scientists must rely on probability to guide understanding.
- The practice of objectivity seeks to limit the influence of natural human bias on the scientific process, from single experiments to entire fields of research.
- Human welfare and society suffer when science is conducted without objective processes in place.

Discussion questions

1. Mythological stories tell us how societies interpreted the world around them in the absence of the scientific method. Use the internet to find a mythological story which interests you and use the process of abductive, inductive, and deductive reasoning to decide if it is likely to be true or false.
2. The Nuremberg salt test shows that some people might benefit from homeopathic treatments due to the placebo effect (the idea that a treatment is good for them despite it having no basis in medical science as we currently understand it). How might you use the scientific method to test this idea?
3. How good is the media at reporting science objectively? Find a news story about a scientific discovery and try to find the research article it is based upon. Do you think the media story reports the finding accurately? How might public understanding of the science change if human bias were to distort the reporting of the research?

Further reading

Imagine the Universe! The Cosmic Distance Scale. Website.

An explanation of Eratosthenes' stick-and-shadow method that he used to calculate the circumference of the Earth from the National Aeronautics and Space Administration—Goddard Space Flight Center.

Magee, B. *The Story of Philosophy* Dorling Kindersley. 2001.

An introduction to European philosophers and the impact they have had on thinking in the modern age.

Women Scientists in the Scientific Revolution—World History Encyclopaedia. Online website.

A look at prominent female scientists who were active during the European scientific revolution (1500–1700) and the challenges they faced in communicating their ideas.

Women in Statistics: Past, Present and Future. Royal Statistical Society. YouTube Video.

A lecture from the Royal Statistical Society which highlights the role of women in building the future of statistics and data sciences.

Urry, L, Cain, M, Wasserman, S, Minorsky, P, Buskirk, R, Reece, J, and Gillen, C. *Inquiry in Action—Interpreting Scientific Papers*, Pearson Education. 2016. ISBN:9780134478616

A two-part guide about how research articles are put together and how to read them.

Kennedy, L. Teaching the Difficult Past of Statistics to Improve the Future. *Shaffer Journal of Statistics and Data Science Education* 32(1), 2024. Doi.org:10.1080/26939169.2023.2224407.

This online open access article describes the way in which statistical principles have emerged from a controversial history to bring benefit and enlightenment to people around the world.

2 WHAT IS 'NORMAL'?

- Populations and how to sample them
- Describing data distributions
- Using the mean and standard deviations
- Why does data vary?
- Size matters!
- Big data

The internet is full of helpful advice about how to feel good and live well. Healthy living apps, for example, tell us we should measure heart rate (pulse) regularly to track daily physical activity. From this, we can manage our personal risk of heart disease and other health problems like type 2 diabetes. As you sit quietly reading this, try measuring your resting pulse. Does it fall somewhere between 60 and 100 beats per minute (bpm)? A healthy living app will tell you that this is within the normal range for a healthy person, but let's think about that statement carefully. Are some resting pulse measurements more normal than others? Is a resting pulse measured closer to either 60 or 100bpm considered 'less normal' than, say, 80bpm, which is in the middle of the range? What about other factors which affect resting pulse like time of day, where and how the measurement was taken, time since your last meal, your sex, racial heritage, age, how much you weigh, and even where you live. Hang on, that's a lot to take in! How are we to make sense of any resting pulse measurement if so many factors can affect it? How do you know if the resting pulse measurement you have just taken is normal? And if everyone else is affected in the same way, then how can any meaningful comparison be made between their measurements and your own? In fact, is there such a thing as 'normal' and can it be measured? This is where **statistics** can help us (Figure 2.1).

Figure 2.1 Everyone is different, so what is normal?

Statistics is the collection, organization, analysis, interpretation, and presentation of data—often followed by drawing conclusions and making decisions based on that data analysis. In fact, statistics underpins every aspect of biology—and every other science.

Populations and how to sample them

Populations—a group of organisms of a single species that live and reproduce in the same place at the same time—are a key biological concept. In statistics a population doesn't have to be a group of living organisms but could be any set of things which share at least one common characteristic. By asking how similar populations might be to one another, we are freed from considering 'normal' and 'not normal'. We move closer to a much more scientific approach where we ask how populations differ and why. In this way, statistics allow us to explore any kind of question where data can be gathered so that comparisons can be made. It gives us a way to explore and understand the physical nature of things, patterns of life, why events happen, and how to predict the likelihood of events in the future.

The first step in any research is to gather accurate information about something you want to investigate. This is used to develop hypotheses which can be tested using the scientific method (Chapter 1). In ideal circumstances, this information would be gathered from every member in a population. However, this isn't usually practical, for example, there may be too many individuals to sample, some may be difficult to access (such as cliff-breeding seabirds (Figure 2.2)), or the true number of individuals in the population might not be known (for example, bacterial growth patterns mean the number of individuals in a population will vary greatly in space and time). But all is not lost. A **sample** provides a representative snapshot of the true population.

Random sampling

For a sample to be representative of the population, any individual must have an equal chance of being captured and this is achieved using a process of **random sampling** (Figure 2.3). For example, you might want to find out how many pupils in a school measure their fitness using an app on their phone, but the school population of 1500 students is too large to ask everyone. The process in Figure 2.3 shows how you would produce a random sample of 30 students by blinded selection or use of a random number generator. In these examples, every student in the school population has an equal chance of being included in the sample.

The size of the sample chosen is very important. The bigger the sample size, the more likely you are to get reliable results. This is because the chance of incorrectly observing something or assigning importance to an

Figure 2.2 This gannet colony on the Yorkshire cliffs, UK, is large, hard to reach, and constantly changing as birds fly to and from their nests. When researchers sample only part of the colony, they capture a moment-in-time that can be used to build a picture of the wider population.

Figure 2.3 A random sample of 30 participants from a population of 1500 can be chosen by drawing names blindly from a pooled list—such as pulling names from a bowl—or by giving everyone in the population a randomly generated number and selecting those whose numbers fall within the first 30.

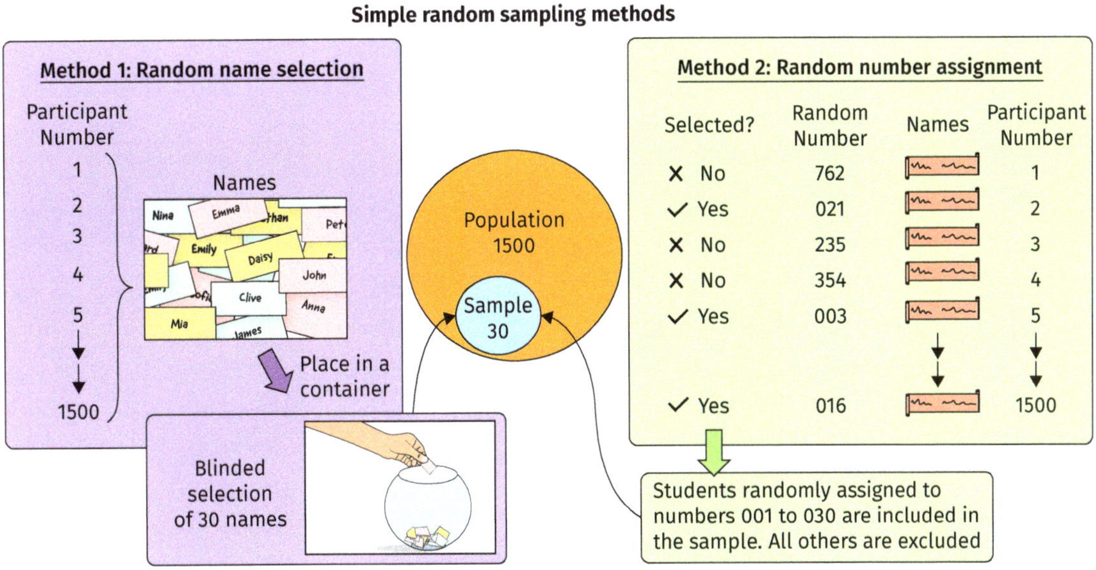

uncommon or irrelevant event ***decreases*** as sample size ***increases***. This is known as the risk of error. Understanding and recording the risk of error helps explain the closeness of data within the sample to the true population, or the study precision.

Describing data distributions

Once we have collected data by sampling a population, we need to use it. For example, how does population sampling help someone understand if their resting pulse is healthy for their age?

In this example, the first step is to understand how resting pulse varies between people who fall within a particular age group.

- Define your age group, for example 16–20 years old.
- Measure the resting pulse using a random sampling method for individuals whose age falls within this range. The larger your sample, the greater your study precision.
- Divide the data into sequential categories, or 'bins', and create a frequency histogram showing number of observations that fall into each bin (Figure 2.4).

The histogram shows that resting pulse in 16–20-year-old students has a shape, or data distribution, which emerges from the number of times an observation falls within each bin. Frequently encountered pulse values fall towards the middle of the distribution and less frequent or rare pulse values

Figure 2.4 Frequency histogram of resting pulse for 16–20-year-old students. The data within each resting pulse bin is given in the table on the right of the figure. The dashed line illustrates the bell-shaped pattern.

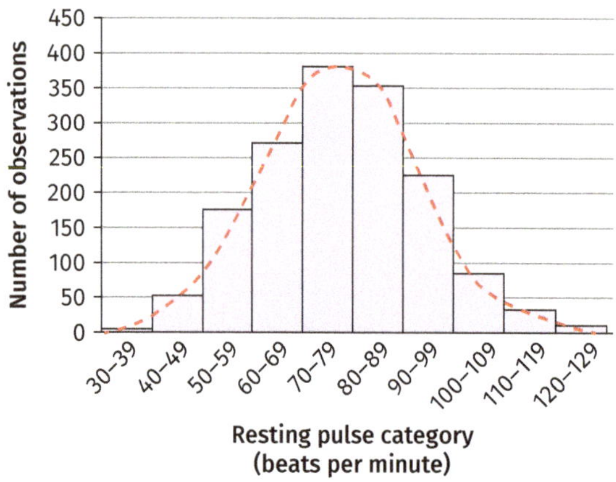

Pulse (beats per minute)	Number of observations
30–39	5
40–49	52
50–59	176
60–69	271
70–79	381
80–89	353
90–99	225
100–109	85
110–119	33
120–129	11
Total sample size (n):	**1592**

fall towards the margins. This is often described as a **normal distribution**, or **bell-shaped curve**, because the data changes more-or-less symmetrically with distance from the mid-point.

It is important in statistics to describe the data distribution, and this is usually done using two types of descriptive statistics:

- **Central tendency** describes the position of the mid-point in a data set.
- Variance or deviation explains how the data changes from the mid-point to the periphery of the data set.

Looking at Figure 2.4, you might notice that the central point of the data is not quite straightforward to find because there are differences between the most frequently observed pulse value, the pulse value which falls in the middle of the range, and the average, or mean, of all pulse values recorded. Figure 2.5 shows how **mode**, **median**, and **mean** provide three measures of central tendency which are commonly used to describe a data distribution.

Mode, median, and mean often produce quite different central tendency values, so how do we decide which to use and when? Mean is the most common measure but should be used with caution if the data set includes values which are very large or small or are far-removed from the middle of the distribution (known as **outliers**). In this case, median provides a better measure of central tendency because it is less influenced by data at the extremes of the distribution. The mode is often used to describe data which falls into non-continuous categories such as biological sex (male or female), type of behaviour (smoker? Yes or No), or blood type (A, B, AB, O).

How do we know if data is normally distributed?

Data does not always fall into the pattern of a bell-shaped curve. Sometimes data scatters towards one end of a range and this causes the mean, median, and mode to separate from each other. This effect is called **skew**. Scientists use normality tests like the **Shapiro–Wilks** or **Kolmogorov–Smirnov** tests to

Figure 2.5 Measures of central tendency and how to find them in the pulse data.

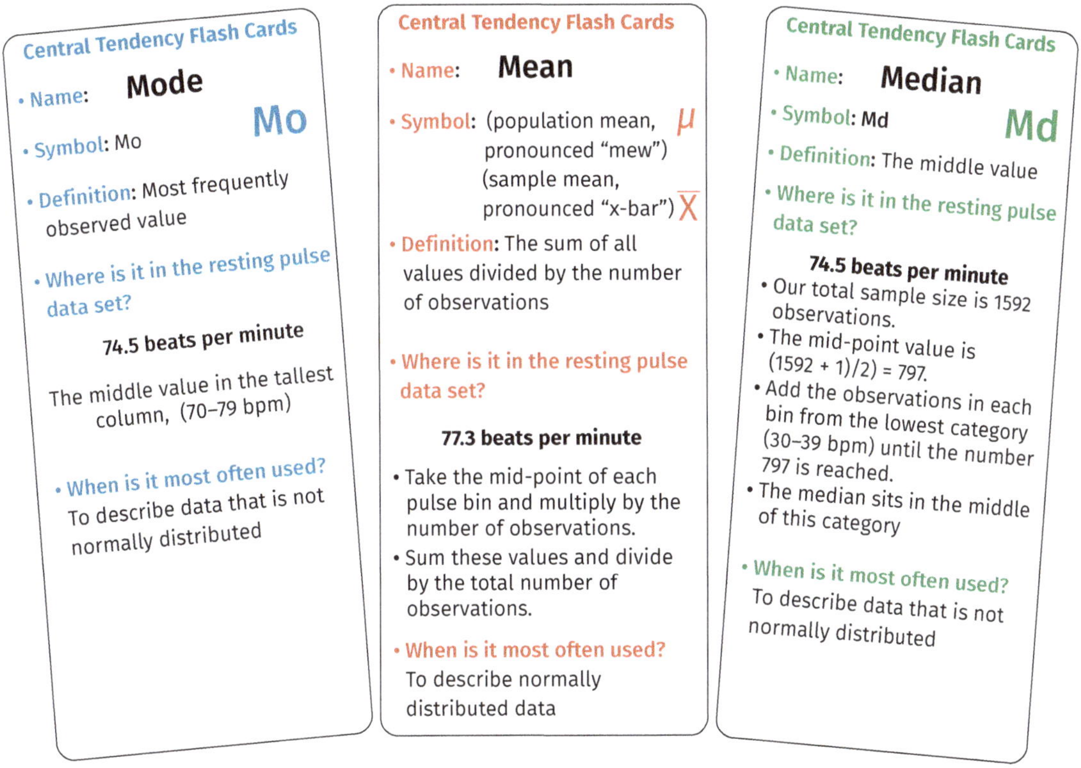

see if data from a sample matches the pattern of a bell-shaped curve or if it is skewed. The results help them decide the best statistical approach for exploring their data (Figure 2.6).

Variance and standard deviation

Sometimes the differences within a population are enormous, but sometimes they are relatively small. We use the concepts of variance and standard deviation to show us how the data falls around the central tendency by describing its range and the distance from each value to the mean.

- Variance describes how far each data point lies from the mean.
- Standard deviation is the square root of the variance and is widely used in science as a standard measure of data spread.

Small standard deviation values tell us that the data clusters closely around the mean with a narrow variance. A large standard deviation indicates that the data vary widely around the mean. You will come across standard deviation in almost any piece of biological research you read about or carry out. Other simple ways to describe data dispersion include dividing distributions into quarter segments (quartiles) or tenths (deciles).

Figure 2.6 The distribution of pedestrians is skewed to each side of the street as they wait to cross at a busy intersection in Tokyo. The graphs below show how the *mean becomes less reliable as a measure of central tendency when data skews in a positive or negative direction.*

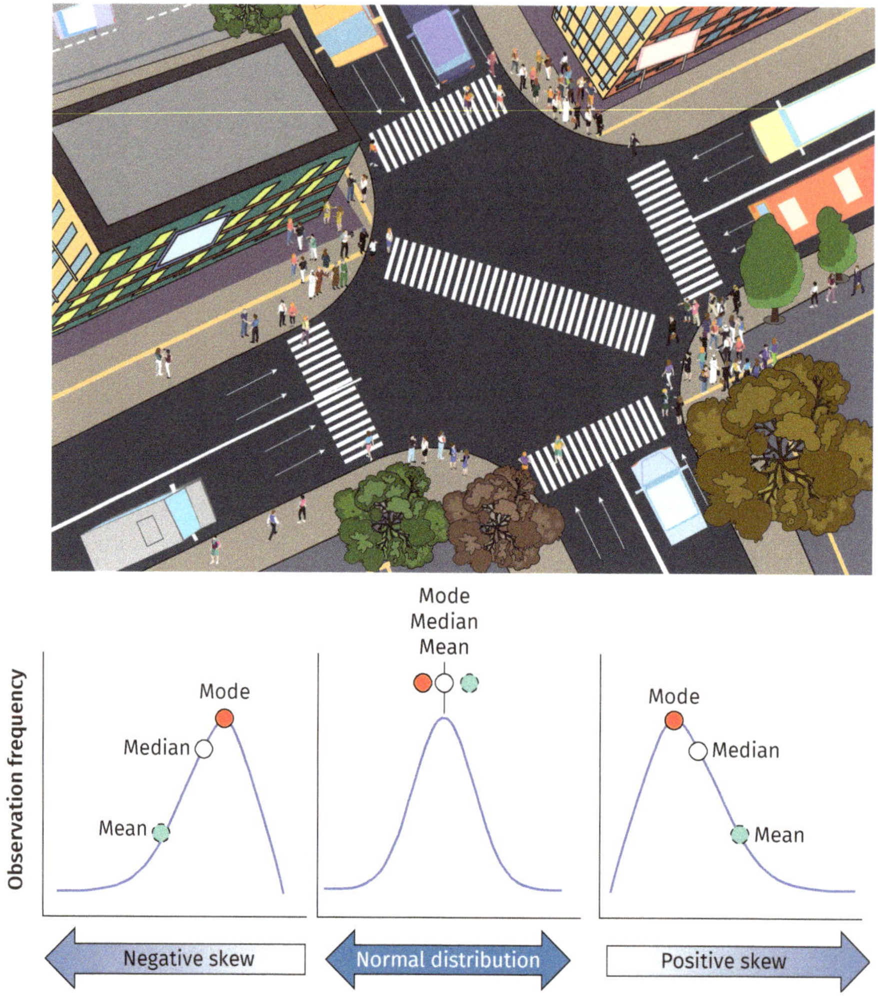

Reading statistical formulas

The flashcards in Figure 2.7 show how statisticians calculate range, variance, and standard deviation in a data set and introduce you to the way that statistical formulas are written. These can sometimes look confusing because they use a mix of letters from the Latin and Greek alphabets, each with its own special meaning.

For example, the Greek symbol for a capital 'S' is sigma, Σ, which is used to tell you to take the sum of the terms that follow it. The next letters will be an instruction to do something and are often written in brackets meaning 'do this together'. For example, in the variance formula in Figure 2.7, Σ is followed by $(x_i - \overline{x})^2$, which means 'take the mean away from each data point, then square it'.

Figure 2.7 Measures of data distribution and what they mean.

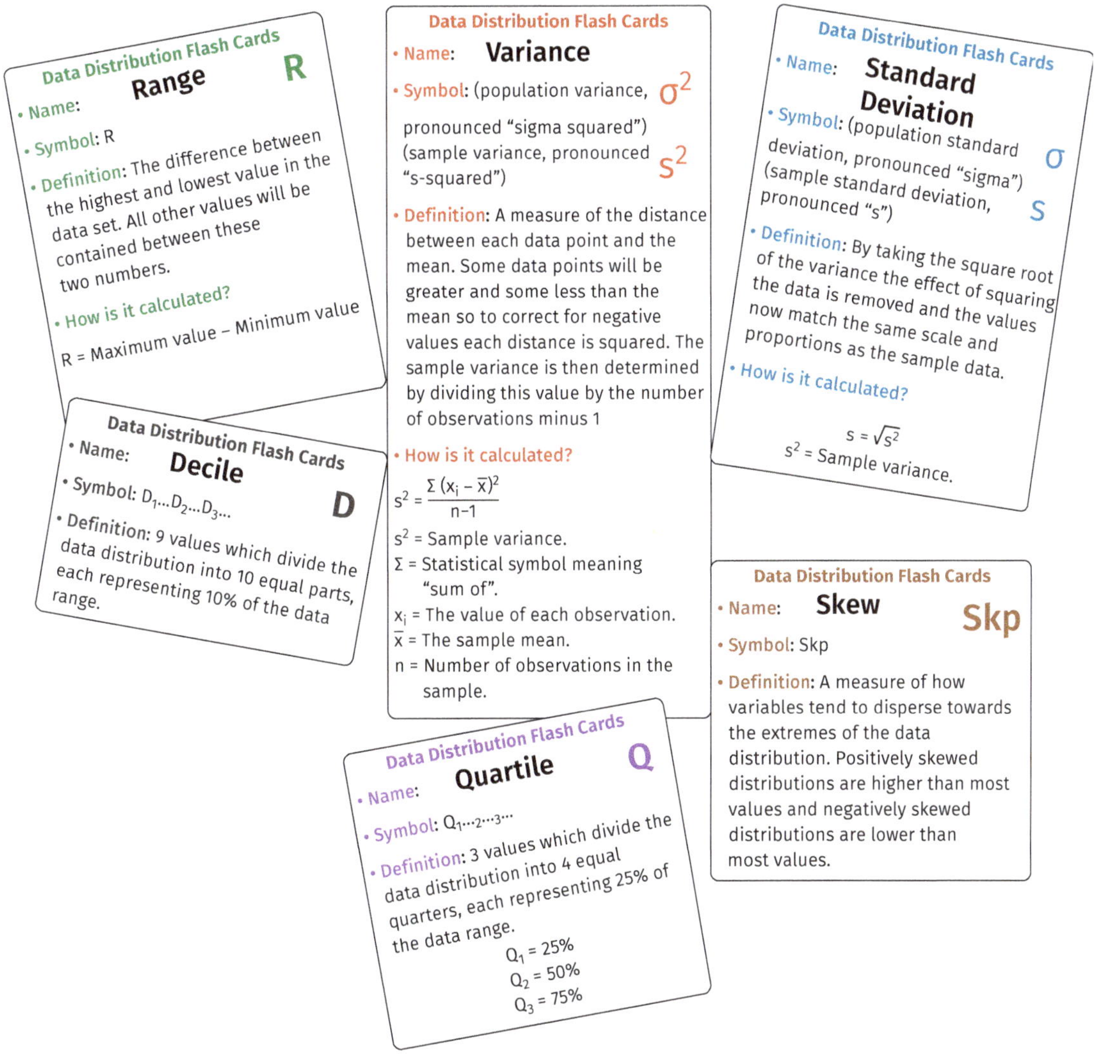

The final term, n – 1, is written under these commands and is outside the bracket. This means 'your last step is to divide the sum of $(x_i - \bar{x})^2$ by the number of observations minus 1'.

Think of each letter as an instruction that leads you through the calculation process. The table includes an explanation of the terms used in each formula.

Using standard deviation to understand data distributions

Biologists use standard deviation to describe the characteristics of a population or data sample. For example, think about a plant conservation study which explores whether wildflower seed mixes sown in field margins

Figure 2.8 Species count in monoculture crops cluster close to the mean with a narrow standard deviation (σ) compared to uncultivated field margins where species are diverse and spread out. Standard deviation can be used to understand how these environments influence each other over time.

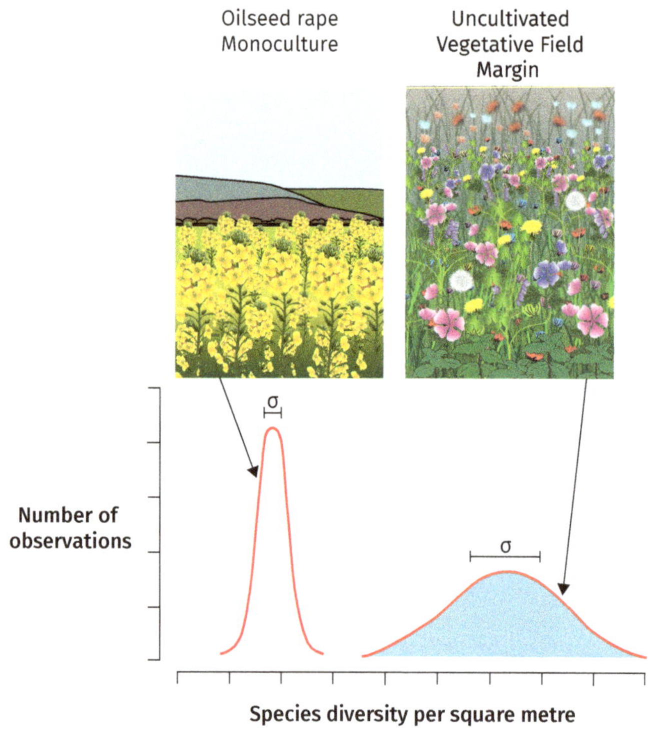

promote crop productivity by encouraging diverse insect pollinators and predators as a natural form of pest control. A researcher might measure the diversity of plant species found in the cultivated part of the field and compare this to the number in the wild-seeded field margin. A count of different plant and insect species per square metre might show a narrow variance and small standard deviation in the cultivated field where the dominant species is the farmed crop. A wider variance and higher standard deviation might well be found in the uncultivated margin where diverse wild plant species encourage insect populations to flourish (Figure 2.8). Conservation researchers use this sort of information to understand how maintaining high plant diversity boosts crop productivity without use of chemical fertilizers or pesticides.

Using the mean and standard deviations

How can we use the mean and standard deviation to find out if a value is common or rare?

Now we have a measure of central tendency, and a standardized method to describe how the data varies, we can begin to make predictions about how data behaves within the distribution. The standard notation for a normal distribution is given in Equation 2.1:

$$X \sim N(\mu,\sigma)$$

where:

X= any sample value chosen at random
~ = varies with
N = normal distribution
μ = population mean
σ = population standard deviation

This tells us that the position of any value within a population normal distribution can be found if we know the population mean and standard deviation.

The scatter of the data for any normal distribution, whether clustered or dispersed, follows a regular pattern where 68.2% of observations fall within one standard deviation of the mean, 95% within two and 99.7% within 3 standard deviations of the mean.

This is known as the **Empirical** or **68–95–99.7 rule** which provides a measurement of likelihood, or **probability density**, that is used to determine if a data point, X, falls within a part of the range that is typical or rare for our sample (Figure 2.9).

Figure 2.9 The Empirical or 68–95–99.7 rule shows how normally distributed data disperses in a regular way either side of the mean. This relationship holds true whether the standard deviation is large or small.

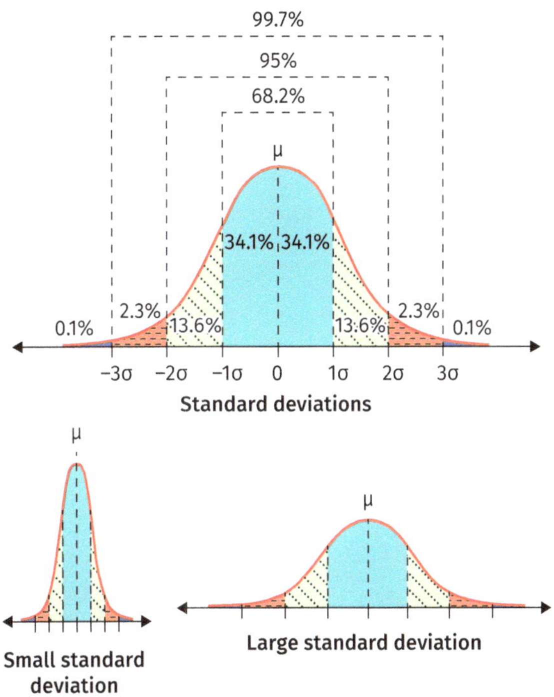

The number of standard deviations that a sample falls from the mean is referred to as the **Z-score** and is described by Equation 2.2:

$$Z = \frac{X - \mu}{\sigma}$$

where:

Z= Z-score (number of standard deviations that a value, X, is from the population mean)
X = value of the sample
μ = population mean
σ = population standard deviation

Since data scatters according to the Empirical rule, the Z-score can be used to work out the probability of any particular observation falling within the normal distribution. Positive Z-score values occur above the mean and negative values fall below.

Let's apply this idea to the resting pulse data in Figure 2.4. A student wants to know if their resting pulse of 85 beats per minute (bpm) is normal. The mean resting pulse value from the data given in the figure is 77 bpm and the standard deviation is 15 bpm. Putting this information into Equation 2.1, we can ask:

> 'What is the probability of selecting a person at random with a resting pulse of 85 bpm or lower if the population mean is 77 bpm with a standard deviation of 15 bpm?'

Figure 2.10 shows mean, standard deviation, and Z-scores and how they describe the normal distribution of resting pulse data from Figure 2.4.

The first step is to calculate the Z-score for X = 85 bpm using Equation 2.2:

$$Z = \frac{85 - 77}{15} = 0.53$$

Figure 2.10 Resting pulse data, standard deviation, and Z-scores align either side of the mean (μ) which is given as the number in the grey box.

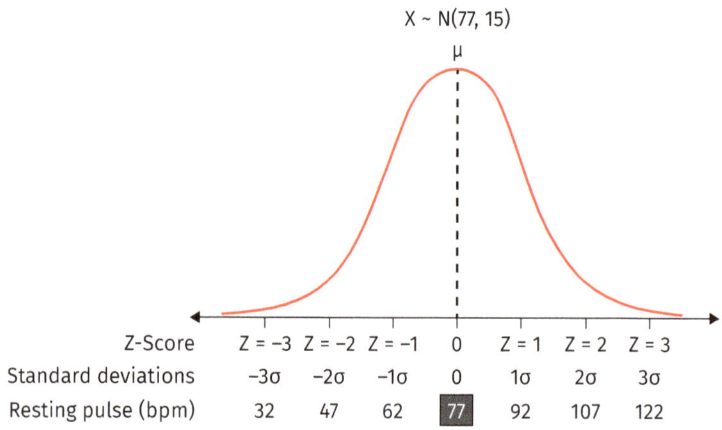

This value tells us that the measured pulse value for this student is located 0.53 standard deviations above the mean. The likelihood of encountering any value up to this Z-score at random is given by the area under the normal distribution curve up to this point. For convenience, we can use **Standard Normal Distribution Tables** to determine the percentage area under a normal distribution curve for any positive or negative Z-score. An extract is shown beneath the graph in Figure 2.11 and an explanation of how to interpret these data follows.

Statistical tables

Statistical tables give researchers a way to look up probability values for specific data types, distributions, or sample sizes. You can think of them as maps that help locate the probability associated with a particular value under a normal curve or other distribution. Different statistical tests often rely on their own specialized tables. While online calculators can perform these calculations instantly, consulting the tables helps you understand where the data comes from.

Figure 2.11 Graph from Figure 2.10 is repeated and includes an extract from the standard Normal Distribution Tables. These tables allow us to find the area under the curve to the left of a given Z-score (0.53 in this example), which represents the probability of observing any value within the purple shaded region. Standard Normal Distribution Tables can be used to determine the area under the curve to the left of a Z-score. This provides a measure of the probability that we will encounter a given value of *X* in the shaded portion of the normal distribution.

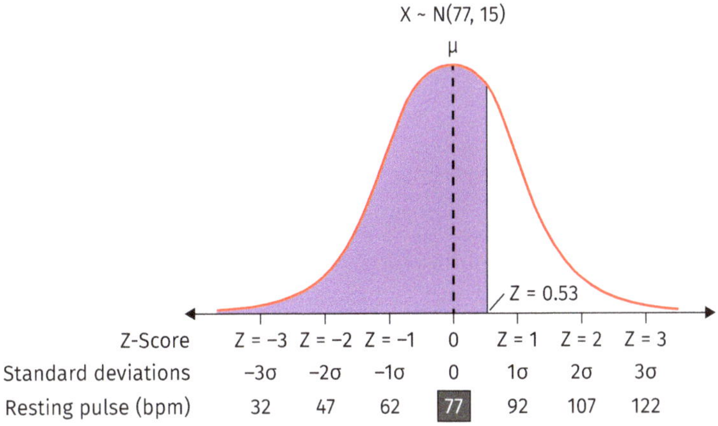

z	0	0.01	0.02	0.03	0.04	0.05	0.06	0.07	0.08	0.09
0	0.5	0.50399	0.50798	0.51197	0.51595	0.51994	0.52392	0.5279	0.53188	0.53586
0.1	0.53983	0.5438	0.54776	0.55172	0.55567	0.55962	0.5636	0.56749	0.57142	0.57535
0.2	0.57926	0.58317	0.58706	0.59095	0.59483	0.59871	0.60257	0.60642	0.61026	0.61409
0.3	0.61791	0.62172	0.62552	0.6293	0.63307	0.63683	0.64058	0.64431	0.64803	0.65173
0.4	0.65542	0.6591	0.66276	0.6664	0.67003	0.67364	0.67724	0.68082	0.68439	0.68793
0.5	0.69146	0.69497	0.69847	0.70194	0.7054	0.70884	0.71226	0.71566	0.71904	0.7224
0.6	0.72575	0.72907	0.73237	0.73565	0.73891	0.74215	0.74537	0.74857	0.75175	0.7549
0.7	0.75804	0.76115	0.76424	0.7673	0.77035	0.77337	0.77637	0.77935	0.7823	0.78524

Using Standard Normal Distribution Tables

The extract of a table shown in Figure 2.11 looks complicated—but using these tables just involves a short series of steps. Here we walk through it using the 0.53 standard deviations of the measured pulse rate from the mean:

- Read down the left-most column to find the calculated value to one decimal place—from our student pulse rate example, this is 0.5.
- Now read across to obtain the second decimal place—in this case 0.03 to give us the combined 0.53 (2 significant figures).
- The whole area under the curve is considered 100%, or 1.000.
- The value of 0.70194 which is returned for the Z-score of 0.53 means that the likelihood of randomly selecting a person within the sample with a resting pulse value up to 85 bpm is 70.2%. Pretty high!

We know the sample population size is 1592 people, and the percentage of the population with a pulse rate up to 85 bpm is 70.2%, so we can work this out in terms of real people.

$$\frac{1592 \times 70.2}{100} = 1117.6 \text{ people}$$

This means that about 1117.6 people out of every 1592 people would have a resting pulse of 85 bpm or lower. That is roughly 1 in 1.4 people—so the student's measurement is typical of the sampled population.

What about the 5 samples in Figure 2.4 which fell between 30 and 39 bpm at the far left-hand side of the sample distribution? Are they unusual? We can apply the same approach by taking the mid-point value of 34.5 bpm for this range. The first step is to calculate the Z-score for $X = 34.5$:

$$Z = \frac{34.5 - 77}{15}$$
$$= -2.83 \text{ (in other words, 2.83 standard deviations below the mean)}$$

The negative Standard Normal Distribution Table in Figure 2.12 returns a value of 0.00233 for these readings. So only 0.23% of the people in the sample population will have a resting pulse less than 34.5 bpm. Since we know the sample population size is 1592 people, this small percentage value corresponds to $(1592 \times 0.23)/100 = 3.66$ people. This is the same as $1592/3.66 = 1$ in 435 people will have a resting pulse of 34.5 bpm or lower.

What is the probability that a person will have a resting pulse of 120 bpm or higher? Following the same process,

$$Z = \frac{120 - 77}{15}$$
$$= 2.87 \text{ (2.87 standard deviations above the mean)}$$

The positive Standard Normal Distribution Table in Figure 2.12 returns a value of 0.00205 indicating that 0.21% of the people in the sample population will have a resting pulse greater than 120 bpm, or 3.34 people, or 1 in 477.

Figure 2.12 Repeat of graph from Figure 2.10 with additions as shown. Positive and negative Z-scores indicate a data distribution which is either greater or less than the mean. Extracts from the Standard Normal Distribution Table shows how positive and negative Z-scores can be used to determine the probability of an observation.

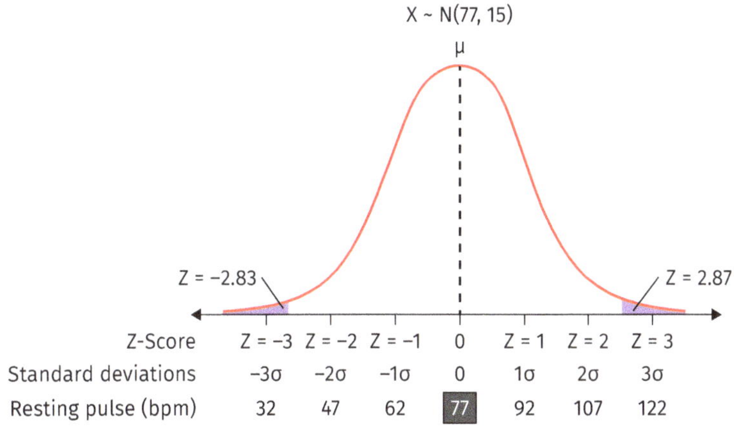

Probability that a value is less than Z										
z	−0.00	−0.01	−0.02	−0.03	−0.04	−0.05	−0.06	−0.07	−0.08	−0.09
−2.9	0.00187	0.00181	0.00175	0.00169	0.00164	0.00159	0.00154	0.00149	0.00144	0.00139
−2.8	0.00256	0.00248	0.0024	0.00233	0.00226	0.00219	0.00212	0.00205	0.00199	0.00193
−2.7	0.00347	0.00336	0.00326	0.00317	0.00307	0.00298	0.00289	0.0028	0.00272	0.00264
−2.6	0.00466	0.00453	0.0044	0.00427	0.00415	0.00402	0.00391	0.00379	0.00368	0.00357
−2.5	0.00621	0.00604	0.00587	0.0057	0.00554	0.00539	0.00523	0.00508	0.00494	0.0048

Probability that a value is greater than Z										
z	0	0.01	0.02	0.03	0.04	0.05	0.06	0.07	0.08	0.09
2.5	0.00621	0.00604	0.00587	0.0057	0.00554	0.00539	0.00523	0.00508	0.00494	0.0048
2.6	0.00466	0.00453	0.0044	0.00427	0.00415	0.00402	0.00391	0.00379	0.00368	0.00357
2.7	0.00347	0.00336	0.00326	0.00317	0.00307	0.00298	0.00289	0.0028	0.00272	0.00264
2.8	0.00256	0.00248	0.0024	0.00233	0.00226	0.00219	0.00212	0.00205	0.00199	0.00193
2.9	0.00187	0.00181	0.00175	0.00169	0.00164	0.00159	0.00154	0.00149	0.00144	0.00139

Case study 2.1

Using standard deviation to study climate change

Measuring climate change is difficult. Temperatures of the Earth's surface can range over 100 °C from the coldest to the hottest places (Figures A and B) with lows and highs occurring at different times from one year to the next. Storage and release of heat occurs at different rates for oceans, land, and atmosphere and we still don't fully understand how these environments naturally affect one another over time.

Figures A & B Temperatures in the Arizona desert (Figure A) and Iceland (Figure B) will vary greatly whatever happens to the global climate—but even a few degrees hotter or colder in either place could have major consequences for the plants and animals which live in them.

A B

These are just some of the reasons why short-term measures of temperature tell us very little about climate trends. Instead, daily temperatures from land and ocean stations are compared to the long-term average at that location over years. This helps to build a picture of temperature variation around the long-term average, which is key to understanding climate change. In the same way that mean and standard deviation allowed us to think about the likelihood of encountering a particular resting heart rate, so these relationships can be used to explore the biology of climate change.

A study by Scott Robeson, geographer and statistician at Indiana University, USA, used historical daily air temperature measurements recorded between 1948 and 1997 at 1062 locations across the USA to build a picture of the relationship between monthly mean maximum air temperature (T_{max}) and monthly standard deviation of T_{max}. By plotting standard deviation as the dependent variable against the T_{max} as the independent variable, scatter plots were produced to show how this relationship behaves over a 49-year period for each month of the year. A line of best fit passed through the data, also known as a **regression line**, shows the trend in temperature variation with monthly maximum temperature. Figure C shows this relationship for one weather station in Canyon City, Colorado, USA.

The downward slope of the regression line for every month shows that standard deviation decreases as monthly T_{max} rises. Remember that small standard deviation values show the data clusters close to the mean. The higher temperatures in each month indicate stable, longer-lasting temperature patterns which have a major impact on local climate. September has the highest variance of all months and so might serve as an alarm bell for a warming climate trend if this were to change. In this month, each 1 °C decrease in standard deviation is associated with a 1.4 °C increase in air temperature. In other words, a warming climate would cause weather patterns in September to behave more like June and July. By monitoring the relationship between monthly T_{max} and monthly standard deviation at different locations over time, we can build a picture of how summer conditions extend further into the autumn months so shortening the winter period.

Figure C Graphs showing the monthly mean maximum air temperature (T_{max}) versus monthly standard deviation of T_{max} for Canyon City, Colorado, over the period 1948–1997. The lines show how standard deviation becomes narrower as monthly temperature rises over time, suggesting a stable warming trend in each month of the year (from Robeson, SM, *Clim. Res.* 22: 205–213, 2002).

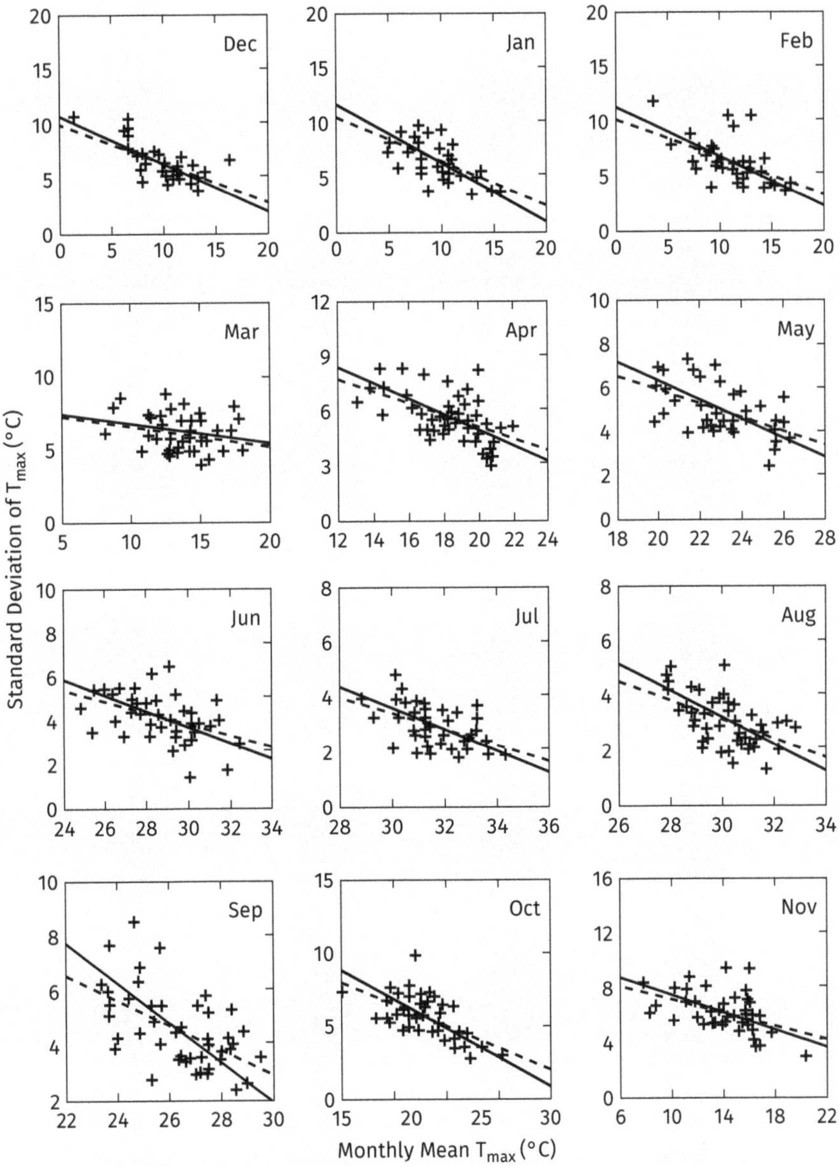

Shorter winters and longer summers have major implications for agriculture and the natural world. For example, reduced winter lake ice cover leads to summer water shortages through sustained evaporation over the year, together with falling water tables due to lack of rain in hotter summers. Cold winters are also an important form of natural pest control and so fewer frosty nights increase insect pest populations, in turn increasing the need for agro-chemicals to control them. Some plants—for example many types of apple trees—need spells of low temperatures in the winter to produce a large crop of fruit the next summer.

Monitoring the relationship between standard deviation and mean maximal temperature over time therefore helps climate scientists understand seasonal variation caused by man-made climate change.

Pause for thought

- Can you think of a way in which changes in the environment have affected you?
- Have you noticed a change in the seasonal availability of produce in the shops, for example, or in the appearance of different birds in your garden, local parks, or school grounds? How might you measure that variation and use it to explain or predict future change?

Why does data vary?

The descriptive statistics we've looked at so far provide a way to think about data patterns in a system—but they don't tell us how or why things vary. It's important to understand how variance affects a system because this has implications for how a statistical problem should be tackled.

There are two causes of variation which cause data distributions to change:

1. **Common-cause variation** arises from naturally occurring effects which are constant, consistent, and predictable. They are characteristic to the system and cannot be avoided. An experiment may include a particular set of controls to manage the effect of this variation on the way the data is collected and analysed. In statistical terms, common-cause variation only ever influences the normal distribution between standard deviations Z+3 and Z−3. Common-cause variation is often referred to as 'noise' caused by things which naturally occur in the system, for example:

   - Cyclical changes in ocean temperature within climate change charts.
   - Any genetic trait, such as albinism, in a population of organisms.
   - The effect of body dimensions on winning times in a 100 m sprint race.

2. **Special-cause variation** is caused by unusual effects which are unpredictable and lie outside the normal distribution of the data (in statistical terms, greater than Z+3 and Z−3). The unpredictability of

special-cause variation can reduce confidence in a data set, since its irregular behaviour makes it hard to manage using standard experimental controls. It often has a particular cause which, if found, will prevent recurrence. Examples of special-cause variation are:

- Human error leading to biased measurement (remember objectivity in Chapter 1?).
- Technical faults in experimental equipment or data collection methods which influence data distribution.
- Unforeseen events which affect the data such as a sudden power cut due to a storm or a change in experimental conditions caused by, for example, a switch in reagents from one supplier to another.

Size matters!

When it comes to statistics, size matters a lot. Common- and special-cause variation show why it can be so difficult to build an accurate picture about something from a single sample of a large population. Standard deviation provides a helpful starting point because the 68-95-99.7 rule tells us how data scatters proportionately around the mean, but this is only a snapshot of how the data behaves in the population. Repeated sampling of the population enhances accuracy by both increasing the overall sample size and allowing for variance to be compared to several sample means. Instead of using standard deviation to describe variance, standard error (SE) explains how variance behaves in relation to sample size. It is calculated using Equation 2.3:

$$SE = \frac{\sigma}{\sqrt{n}}$$

where:
σ = standard deviation of sample meansn
n = sample size

As more observations are made from several samples of the population, so standard error decreases; in other words, the greater the number of samples, the more accurate the measure of true natural variation of the population. SE is used to produce a margin of error estimate for the mean and is normally reported as range of data values in which the true mean is likely to be found. This range, known as a confidence interval (CI), gives a measure of study precision. It narrows as the study size increases and so more information about population variance is gathered. CI are normally reported as an upper and lower value around the mean with a percentage value to indicate study precision. You may have seen these as whisker marks on graphs. For example, a 90% CI with a range of plus or minus 3 indicates that the true mean will fall somewhere inside this range on 90 out of every 100 occasions it is measured. Confidence intervals of 95% and 99% indicate increasing study precision.

Confidence intervals are calculated using SE values together with pre-set Z-scores for the desired level of precision using Equation 2.4:

$$\textbf{Confidence Interval (CI)} = \textbf{Z} - \textbf{score} \,.\, \textbf{Standard Error}$$

Precision	Z-Score
90%	1.645
95%	1.96
99%	2.575

In statistical equations, the multiply symbol is replaced with a dot to avoid confusion with the term 'x' which is used to indicate a data variable.

Equation 2.4 is easily applied to continuous data where we know the standard error and the level of precision we wish to apply. As sample size increases, so more information about the true population variance is obtained, the confidence interval range narrows, and the precision of the study increases (Figure 2.13).

How are confidence intervals assigned to categorical data such as surveys? Suppose we want to know how many people attend gym regularly for our resting pulse study in Figure 2.4. To explore the effect of sample size on the precision of our data, we could ask a sample of 10, 30, or 100 people if they agree or disagree with the following statement:

'I exercise at the gym twice or more every week'.

The results of our survey are shown in Table 2.1, and we use variations of Equations 2.3 and 2.4 to calculate standard error and the confidence interval range. To begin with, let's calculate the 95% confidence intervals for a sample size of 10 people. Here, 4 people agreed with the statement and 6 did not. The first step is to calculate the proportion of the sample who agree. Statisticians call this p-hat ($\hat{p}$):

$\hat{p} = 4/10 = 0.400.$

Figure 2.13 Study precision increases with sample size as more information is gathered about variation. For 95% confidence intervals, the true value of the mean will lie between the boundaries of the whisker bars in 95 of 100 occasions that it is measured.

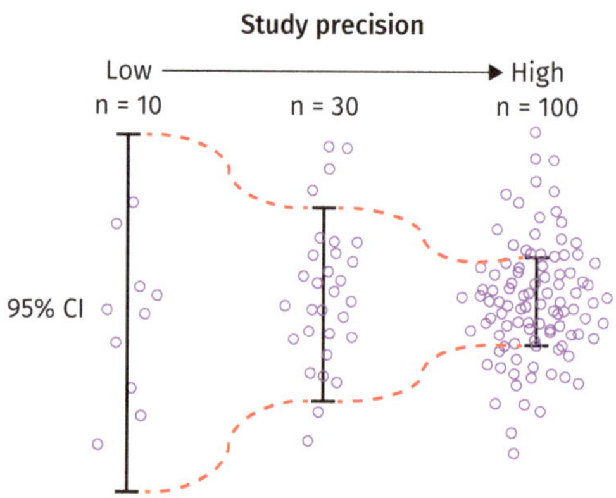

Table 2.1 Survey responses to people who were asked if they agree with the following statement: '***I exercise at the gym twice or more every week***'.

Do you agree with the statement?	Number of Responses	Sample Size
Yes	4	10
No	6	
Yes	9	30
No	21	
Yes	28	100
No	72	

The proportion who disagree is known as q-hat ($\hat{q}$):

$\hat{q} = 1 - \hat{p} = 0.600.$

The standard error of the estimate is calculated as

$$\begin{aligned} \text{Standard error} &= \sqrt{(\hat{p} \times \hat{q})/n} \\ &= \sqrt{(0.4 \times 0.6)/10} \\ &= 0.155 \end{aligned}$$

Using Equation 2.4 together with the 95% Z-score value, the 95% confidence interval is calculated as:

$$\begin{aligned} \text{CI} &= 95\% \text{ Confidence Z-score} \times \text{SE} \\ &= 1.96 \times 0.155 \\ &= \underline{0.304} \end{aligned}$$

The 95% CI range is calculated as $\hat{p}$ plus or minus CI, or,

$0.400 + 0.304 = 0.704$
and
$0.400 - 0.304 = 0.096$

Multiplying $\hat{p}$ and the upper and lower confidence numbers by 100 gives each number as a percentage, so the survey result from a sample size of 10 estimates that 40% of people attend the gym twice or more every week but the true number will fall somewhere between 9.6% and 70.4% on 95 occasions out of every 100 times that 10 people are asked this question.

If you repeat this approach for the sample sizes of 30 and 100 respondents to the survey, then the 95% confidence interval narrows as sample size increases because we have more information about the true population variance.

The survey of 30 people produced 9 who agreed with the statement or 30% with a margin of error of 13.6–46.4. A survey of 100 people produced 28 who agreed with the comment (28%) with a margin of error of 19.2–36.8.

Figure 2.14 The greater the number of people sampled in a survey, so the greater the precision of the study. The number of positive responses to the survey question in Table 2.1 is shown as a black circle with 95% confidence intervals (95% CI) given by the vertical black whiskers.

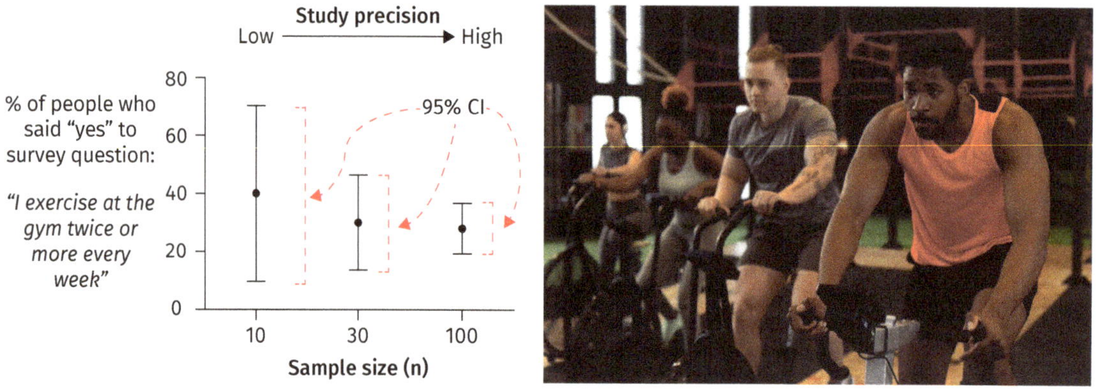

Figure 2.14 shows the number of positive responses plus and minus the 95% confidence interval range. Notice how the number of people responding positively to the survey is greatest in the smallest sample sizes but stabilizes at a lower number between the middle and largest sample sizes as the 95% CI values narrow. As study precision increases with larger sample size, confidence increases that the true result of the survey will lie within this range.

Big data

'Big data' refers to extremely large and complex data sets that are difficult to manage using traditional tools. These data allow researchers to identify intricate patterns and relationships with a high degree of precision. One of the earliest uses of big data were reference data sets that were built to study normal or abnormal health traits over time. These are found in national statistics databases, scientific studies, and on-line resources known as biobanks. Figure 2.15 shows a reference database from the US National Health and Nutrition Examination Survey (1999–2008) which reports resting pulse rate from 47,610 males and females from birth to 80 years of age.

Two methods are used to show how the data distribution changes over time. Firstly, look at the graphs for males and females. The change in median resting pulse is shown as a dark line in the middle of two shaded ribbons which show the data distribution. These indicate percentile and quartile ranges within the data and help the doctor to quickly determine if a patient's resting pulse is typical or cause for concern. Percentiles are determined by converting the data distribution to 100%. The distribution is separated into quartiles using markers at 25 (lower quartile), 50 (median), and 75% (upper quartile). Values which fall 25% above and below the median lie between the 25th and 75th percentiles, also known as the inter-quartile range. This means they are within one standard deviation of the mean (see the Empirical rule). All values that are distributed 45% above and below the median lie within the 5th and 95th percentiles which range outwards to two

Figure 2.15 Quartiles in a reference data set help doctors to know if a measurement falls inside a typical range of values. In this example, quartile resting pulse for men and women from birth to 80 years of age are shown as graphs (top). Complex information about central tendency and data distribution from one of these graphs can be summarized in a box plot (bottom). Data from the US National Health and Nutrition Examination Survey, 1999–2008.

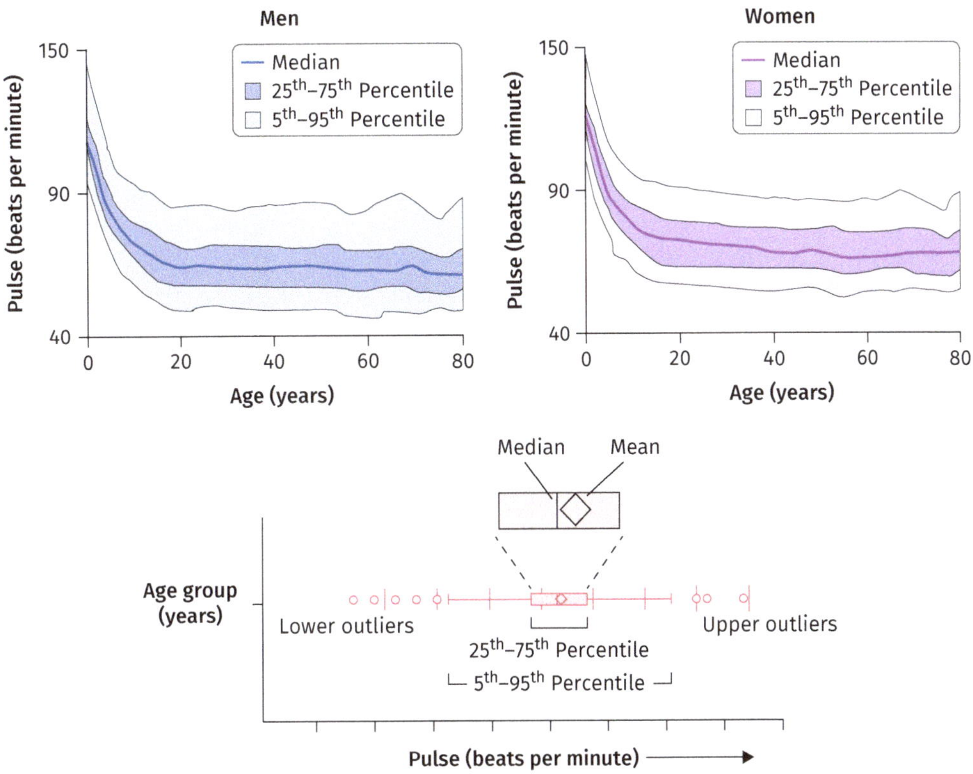

standard deviations. A doctor might use this scale to determine if a patient's resting pulse was typical (within the 25th-75th percentiles) or a cause for concern (towards the limits of the 5th-95th percentiles).

Ways of representing data

Box plots (Figure 2.15) are used to summarize data distributions so that it is easier to see measures of central tendency and variance. In this example, the position of the mean and median are given as a diamond and line symbol. The length of the box shows the inter-quartile range, and the whiskers give the 5th and 95th percentile range. Data points outside this range are called outliers. They might be caused by special-cause variation and cannot be excluded without explanation.

Using big data to understand risk factors

Reference data sets help scientists to understand the biological changes that occur when people are affected by life-changing events such as ageing, disease, or lifestyle choices. They act as a kind of measurement tool for determining whether a condition is likely to get better or worse over time.

In statistics, this is referred to as **risk**, which is given in **Equation 2.5** as:

$$\text{Risk} = \frac{\text{Number of events experienced in a sample}}{\text{Number of participants in the sample}}$$

The event might be any one of the life-changing events mentioned above. It's important to remember that statistical risk is a measure of chance and can be used to measure positive as well as negative outcomes. Biological changes which are caused by an event, but which do not alter its course, are known as **risk markers** and indicate likelihood that a condition will occur. For example, the tendency of glucose to stick to red blood cells in response to excess blood sugar raises the abundance of sugary (glycated) haemoglobin, or HbA1C, and is used as a risk marker of diabetes. Biological changes which cause the event and influence its course over time are known as **risk factors.** These are distinguished from markers because treatments which are aimed at them will alter the course of the disease in one way or another. For example, suntan lotion helps to prevent sunburn which is a risk factor for melanoma.

Scientists often want to know if an event is more likely to occur in one group compared to another, or **relative risk (RR)**. We'll show how relative risk is calculated using the data in Table 2.2. Scientists looked at whether statins (drugs commonly used to lower the level of cholesterol in the blood) change the risk of having a stroke. In the investigation, some patients were given statins and others were given a **placebo**, a pill with no active ingredients.

We can place these data into a matrix which shows the number of people in each test treatment group who experienced stroke. The relative risk (RR) is calculated as shown in Figure 2.16.

If RR is <1 then the risk of stroke is greater for the placebo group compared to the statin group. If RR is >1 then it is the statin group who have the higher risk. This result tells us that the risk of having a stroke whilst taking statins is 25% lower than if not taking statins.

Not all studies lend themselves to calculation of risk. For example, case control studies compare two groups of people to see if there is an association between an exposure to disease and a health outcome. The cases are the number of people who have the disease, and the controls are similar types of people who did not. In this situation strength of association is measured using an **odds ratio** calculated in Equation 2.6 as:

$$\text{Odds} = \frac{\text{Number of people experiencing an event}}{\text{Number of people who did not experience the same event}}$$

Table 2.2 Number of patients on statin or placebo treatments who experience stroke.

Treatment	Number of patients experiencing stroke	Number of patients with no symptoms of stroke
Statins	50	120
Placebo	40	62

Figure 2.16 Calculation of relative risk, or risk ratio (RR), of stroke for patients taking statins compared to a placebo group.

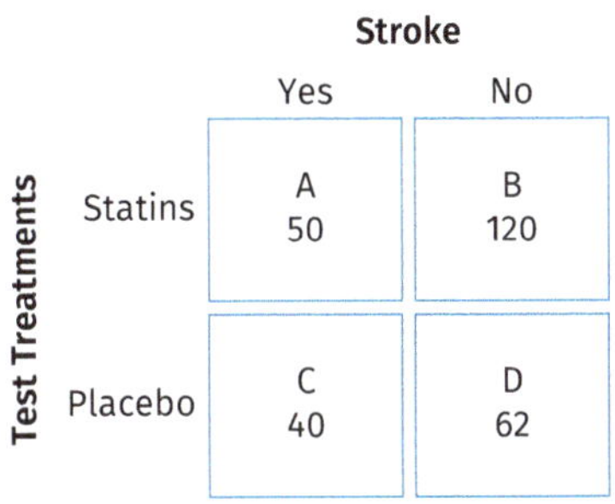

$$\text{Risk Ratio (RR)} = \frac{A/(A+B)}{C/(C+D)}$$

$$RR = \frac{50/170}{40/102}$$

$$RR = \underline{0.75}$$

An odds ratio (OR) is calculated in case-control studies by taking the ratio of the odds for the case over those for the control. We can use this to ask if exposure to an intervention such as a vaccine has any effect on the odds of developing a particular disease. For example, the BCG vaccine is given to protect people against tuberculosis (TB). By collecting health data we can use the odds ratio to determine if the BCG effectively protects against developing TB in later life (Table 2.3).

By placing these data into a matrix, the odds ratio (OR) is calculated as shown in Figure 2.17.

The odds ratio of 0.59 tells us that the odds of developing tuberculosis were 41% lower in the vaccinated group than in the unvaccinated group.

The Office for National Statistics publishes data sets and statistics about the economy, population and wellbeing of society in the UK. It gathers most of this data using government surveys such as the national census, conducted every 10 years. The data is available to the public so anyone can explore national data for different measures. For example, the People, Population and Community page contains databases about health and well-being using measures of central tendency and standard deviation to explain their data.

Table 2.3 Number of patients receiving the BCG vaccine who develop tuberculosis and those who do not.

BCG Vaccine	Number of people who develop tuberculosis	Number of people who do not develop tuberculosis
Yes	99	464
No	110	306

Figure 2.17 Calculation of odds ratio (OR) in a case–control study examining BCG vaccination influence on the incidence of tuberculosis.

Intervention BCG Vaccine	**Disease outcome** Tuberculosis: Yes	Tuberculosis: No
Yes	A 99	B 464
No	C 110	D 306

$$\text{Odds Ratio (OR)} = \frac{A/B}{C/D}$$

$$\text{OR} = \frac{99/464}{110/306}$$

$$\text{OR} = \underline{0.59}$$

Case study 2.2

The Framingham Risk Score

Up to the end of World War II, death from heart disease was considered a natural outcome of a life well led and was not widely seen as a threat to national health. The death of the war-time American President, Franklin D Roosevelt, changed this thinking. Roosevelt paid little attention to his heart health and tended to put occasional headaches and tiredness down to the strains of his Presidential office rather than lack of exercise or an unhealthy diet.

Shortly after attending the Yalta conference in 1945 with Stalin and Churchill, which decided the post-war administration of Germany and founded the United Nations (Figure A), Roosevelt died of a stroke at the age of 63. It is reported that his blood pressure was 300/190mmHg, more than double the level for a healthy person. Roosevelt was a national hero, and his loss was deeply felt in the USA. His successor, President Harry Truman, created the National Heart Act which established the National Heart Institute (now the National Heart, Lung and Blood Institute) and allocated government funding to discover the causes and cures for heart and vascular disease.

What causes heart disease?

One of the first tasks of the Institute was to understand how heart disease occurs in a typical population. On the outskirts of the city of Boston, USA, sits a medium-sized town called Framingham. Framingham's most important feature is that it is 'typical'. At the time of Roosevelt's death, it was a factory town

Figure A Franklin D. Roosevelt sits between Winston Churchill (left) and Joseph Stalin (right) at the Yalta Conference, February 1945. Churchill's physician noted Roosevelt appeared unwell. A few weeks later, on the 12th of April 1945, he was to suffer a fatal stroke caused by high blood pressure.

with low-to-middle-income residents of mostly European origin. The people living there were thought to make similar lifestyle choices and experience much the same environmental influences as other industrial towns in the USA. This common-ness made Framingham a suitable sample population for researching risk markers and risk factors for disease (Figure B). Framingham became the centre of one of the largest epidemiological studies ever conducted. Epidemiology examines patterns of disease—who develops a particular condition, where they live, and the factors that may cause it or increase the risk of it occurring.

In 1948 the US government commissioned the **Framingham Heart Study** which aimed to find out if suspected risk factors such as diet, exercise, age, sex, and genetics might be linked to heart disease. They hoped to find healthcare interventions which could extend healthy lifespan and provide doctors with the risk marker information they need to help people at risk of heart disease through their lives.

The original 1948 study consisted of 5209 men and women aged between 30 and 62 years old. It was realized that risk factors could affect heart function at different rates and different ages, so the selected age range helped researchers to see how risk markers and factors align with heart disease over a lifetime. Some patients were assigned a daily dose of common medicines such as aspirin to determine if these might lower the risk of heart disease. Others gave permission for their family history to be traced from one generation to the next to determine the genetic factors that cause heart disease.

Researchers found that certain factors seemed to follow the development of heart disease better than others. For example, blood pressure during the

Figure B The Framingham Heart Study used a sample of 5209 men and women aged 30–62 to study risk factors for heart disease in the USA. The present-day population of the USA is over 62 thousand times this original sample size, but the study findings remain as relevant today as they were over 60 years ago.

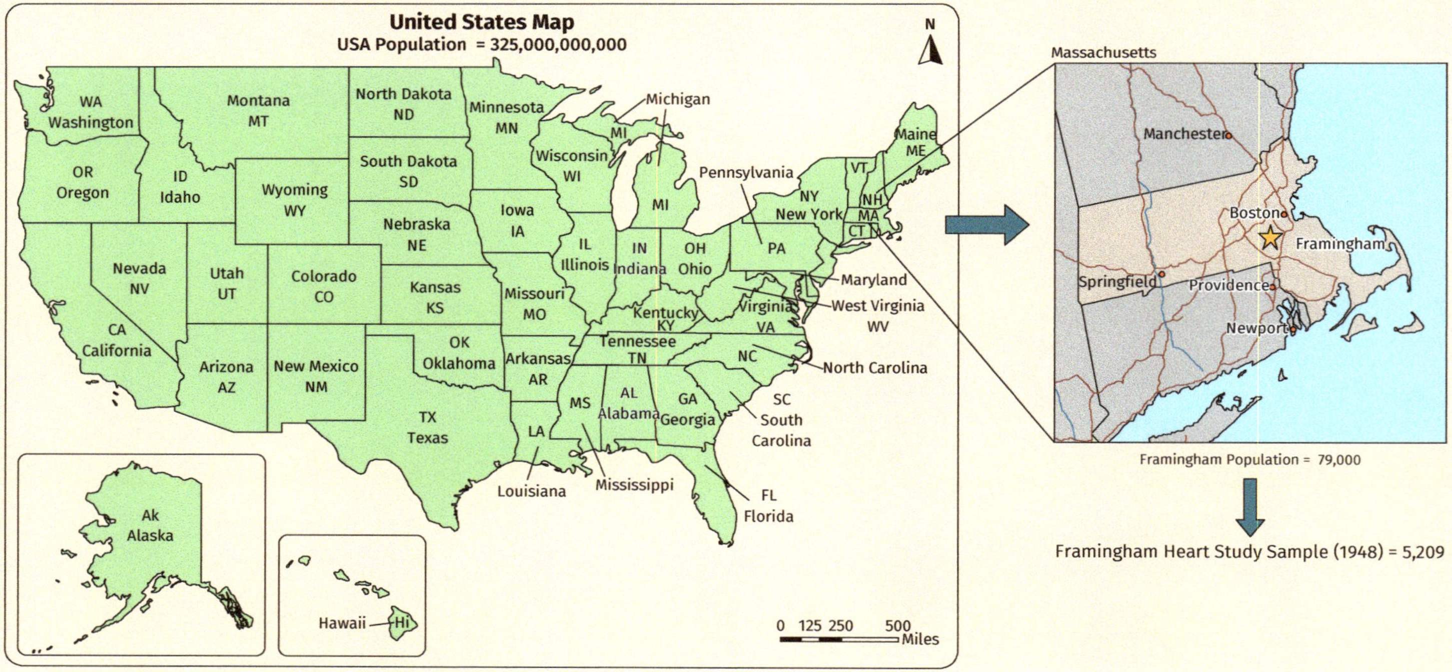

systolic contracting phase of the heart was found to be a much better predictor of heart disease and stroke compared to diastolic pressure during relaxation. Treatments which lowered systolic blood pressure, such as the beta-blocker, propranolol, reduced the risk of heart failure. As a result, in 1977, government advisors recommended that systolic blood pressure should be adopted as a risk factor for cardiovascular disease whereas diastolic blood pressure remained a risk marker.

Identified risks

The risk factor concept laid a vital foundation for developing accurate scoring systems using combined risk factor scores to provide an overall measure of disease risk. The **Framingham Risk Score** looks at five risk factors to predict chronic heart disease risk in middle-aged patients:

- Age
- Blood cholesterol
- Blood pressure
- History of diabetes
- History of smoking.

Each factor is assigned a score to reflect its contribution to the risk of developing heart disease for men and women. When all scores are added together, the combined, or multivariate, risk of developing heart disease within the next 10 years is calculated. This allows doctors to predict whether a patient is likely to develop symptoms of heart disease within 10 years.

Figure C The Framingham study has produced easy-to-use risk charts that allow the danger of developing cardiovascular disease to be calculated from risk factors such as blood pressure. Today, these charts are used by healthcare professionals around the world.

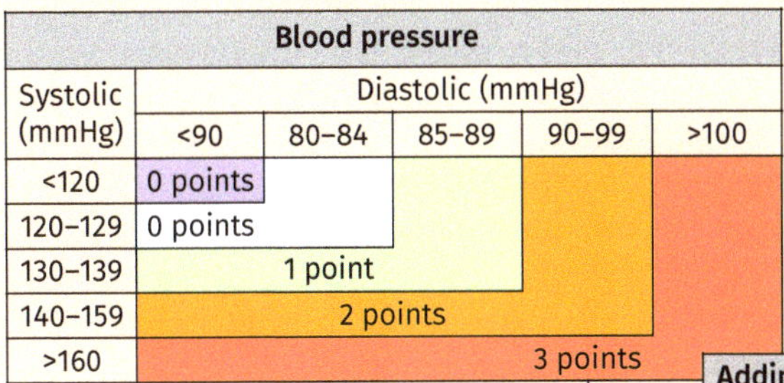

Adding up the points	
Age	2
Cholesterol	2
Blood Pressure	3
Diabetes	1
Smoker	0
Total	8

Total points	Risk of cardiovascular disease (%)
−1	2
0	3
1	3
2	4
3	5
4	7
5	8
6	10
7	13
8	16
9	20
10	25
11	31
12	37
13	47
14	53

The Framingham Risk Score is continually updated to improve accuracy. An approach known as meta-analysis (see Chapter 5), which combines the results of many separate studies to reveal overall patterns, showed that the original version of the scoring system overestimated the risk of developing cardiovascular disease. Nowadays, studies such as these are used to calibrate the scoring system to enhance accuracy.

Now in its seventh decade, the Framingham Heart Study is one of the most rigorous epidemiological studies of its type and, together with the British Doctors Study (Chapter 5), helped to establish heart and circulatory disease links between smoking, obesity, and exercise.

? Pause for thought

- Health service providers like the UK National Health Service provide on-line body mass index (BMI) calculators which give health guidance by comparing your body mass to population data derived from national databases. Can you find a reliable on-line tool to calculate your BMI? How do your results compare to the population?

Mining of massive data sets in the 21st century

In this chapter we have looked at how data distributions are described and explored ways in which the data can be found to be reliable or not. We've established that study size is important for confidence and that the way in which experiments are set up can limit special-cause variance. Where does statistics go from here?

Humans are a naturally curious and social species. We share information daily with others through mobile phone apps, social media, email, and a host of other electronic formats which, under the right circumstances, can be used to explore statistical trends with a level of resolution that has never previously been possible. This has established the new discipline of **massive data mining.** This uses **artificial intelligence** to extract information from ethically approved sources and store it as big data repositories. **Machine learning** algorithms help to identify and remove common and special causes of variation, and **deep learning** algorithms, whose origins were developed from our understanding of neural networks, extract progressively more useful models from the data. The outcome from this process is to develop models which can be used with a high level of resolving power to predict trends in any aspect of science from quantum physics to human health and behaviour (Figure 2.18).

There are, however, major ethical implications about the right of the individual to retain ownership of their personal data and how this is used by organizations and nations. In the European Union, this is governed by **General Data Protection Regulations** which form part of the Charter of Fundamental

Figure 2.18 Massive data mining involves use of artificial intelligence and machine learning to collect, store, and analyse large data sets known as 'big data'. Deep learning computer algorithms help to analyse the data to make predictions which can be used to understand trends in science.

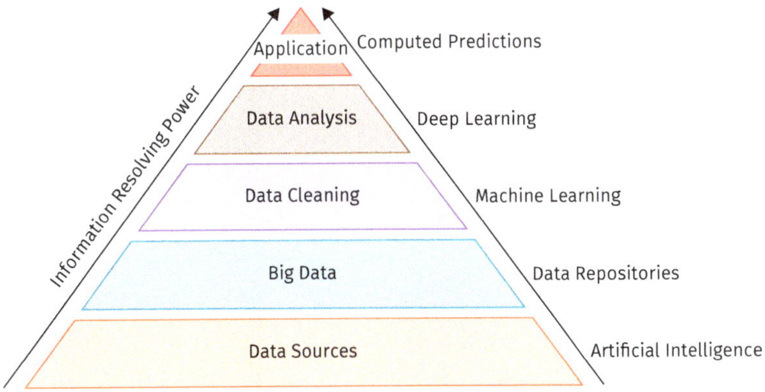

Rights of the European Union and which is adopted by the United Kingdom and numerous other nations around the world.

This chapter began with a question: how do you know if your resting pulse is normal for your age? So, what is the answer? A massive data-mining study published in the journal *PLoS ONE* followed 92,457 regular wearers of wrist heart rate trackers to look at daily resting pulse and its association with age, body mass index, sex, and sleep quality over the course of a year (Giorgio Quer and others, *PLoS ONE* volume 15, e-publication e0227709, 2022). It measured 32 million daily pulse values with enough statistical precision to produce a valid conclusion. The study concluded that each person has a daily resting pulse which is normal for them, but which can differ from another person by as much as 70 bpm. This means that although we might expect resting pulse values to vary considerably in a population, individual resting pulse is a valuable early risk marker of disease. So, in answer to the question, 'is your resting pulse normal?', the precise answer from a massive data-mining study is: 'Yes, your resting pulse is normal to you, but may differ by quite a bit to others around you. A change in your daily resting pulse might indicate a change in your health. This change may not exceed what is normal for other people'.

Chapter summary

- Measuring what is normal and what is not normal in a population is not easy because many factors influence the behaviour of data in a sample.
- Variance and standard deviation help us to describe the distribution of data around a measure of central tendency.
- The Empirical or 68–95–99.7 rule of standard deviation can be used to determine if a value is common or rare in a normally distributed data sample.

- Scientists can use standard deviation to study patterns of variance in changing systems, such as influence of climate change or species distribution.
- The size of a sampled population determines the precision of the measurements a scientist is making.
- Big data studies conducted over several years have been used to create massive data sets which help scientists and doctors identify and explain the risk or odds of developing a disease.
- Data mining of massive data sets is now being used to develop beneficial strategies for predicting risk of disease for individuals.
- The ethics of data mining are being considered by governments around the globe.

Discussion questions

1. A study from the University of Kentucky School of Medicine found that people who carry two copies of a gene called AVI cannot taste the bitterness of certain vegetables. In contrast, individuals with two copies of another gene, known as PAV, are "super-tasters" and experience these same vegetables as intensely bitter. How would you design a randomized study and questionnaire to measure the odds of tasting the bitterness of vegetables? Can you use this approach to make predictions about gene traits using Mendel's laws of genetics?
2. Case Study 2.1 shows how standard deviation can be used to measure changes in climate over time. Can you think of a way in which you might use reliable media reports from the internet to measure changes in the frequency of unusual weather events such as hurricanes, tornadoes, droughts, snowstorms, over the last 10 years? What might the similarities and differences in the frequencies of these events tell you about random occurrence versus climate change trends?
3. As genetically unique individuals, do you think it is reasonable for health providers to use population data to provide personal health advice? How are such databases already being used to develop strategies for personalized medicine? And where might they go in the future?

Further reading

More or Less: Behind the Stats. A BBC Radio 4 and BBC Sounds Podcast.

A weekly radio programme which takes a look at statistics in the popular media and how they help us to understand the world around us.

A brief history of big data everyone should read. World Economic Forum website. 15 February 2015. Online article.

A brief history of big data by Bernard Marr. This news and views article from the World Economic Forum provides a thought-provoking timeline of data collection from the advent of tally sticks and the abacus through to the personal computer.

'*People, population and community.*' Office for National Statistics. Government of UK. Website. See more with a quick search online.

The People, population and community section of the website of the Office for National Statistics is where it publishes data sets and statistics about the economy, population and wellbeing of society in the UK. You can explore their website to see how they use measures of central tendency and standard deviation to explain their data.

Levy, D. (M.D.) and Brink, S. *A Change of Heart: How the Framingham Heart Study Helped Unravel the Mysteries of Cardiovascular Disease.* Knopf. 2005. ISBN:9780375412752.

The story of how the Framingham Study worked as a partnership between the US National Institute of Health and the people of Framingham to shape our understanding of disease risk.

Peck, R, Olsen, C, and Devore, J. *Introduction to Statistics and Data Analysis*, 5th Edition. Brooks Cole. 2015. ISBN 9781305267244.

An excellent text for advancing your understanding of statistical principles.

3 A ROUTE-MAP FOR DISCOVERY

Orienteering is a popular adventure sport where runners navigate between checkpoints in the fastest time across unfamiliar countryside (Figure 3.1). There are no set rules about routes but careful decisions are needed along the way to avoid dead-ends and obstacles. It's a real test of navigational skills and a lot of fun! Statistical tests become most useful when we use them to explore relationships between two or more things where, much like a compass on an orienteering course, they work as a tool to help scientists navigate the unknown. The aim is to build a picture of **causation**: the understanding of whether one thing affects another or if different things have the same origin. Without statistical probability measurements, there would be no path for understanding causation. When thought of like this, statistics are the route-map for discovery.

Figure 3.1 An orienteer checks a control point with a map and a compass. Statistics works in a similar way to help scientists navigate the unknown.

A route-map for discovery

On any road-trip, wrong turns, roadworks, and diversions affect the direction and speed of travel, and so advanced planning helps avoid pitfalls. The same is true in any scientific investigation. In statistics, the roadblocks take the form of false positives and negatives known as type I and II errors.

- A **type I error** is often called a false positive and occurs when a researcher rejects the null hypothesis even though it is actually true. In other words, the statistical test suggests there is an effect when, in reality, no such effect exists.
- A **type II error** is often called a false negative and happens when a study fails to detect a real effect. The researcher accepts the null hypothesis even though it is actually false.

An experimental plan with statistical checkpoints helps to reduce the sources of error and provides scientists with a way to explain how they might affect a study (Figure 3.2).

Figure 3.2 Checkpoints on the hypothesis testing route-map.

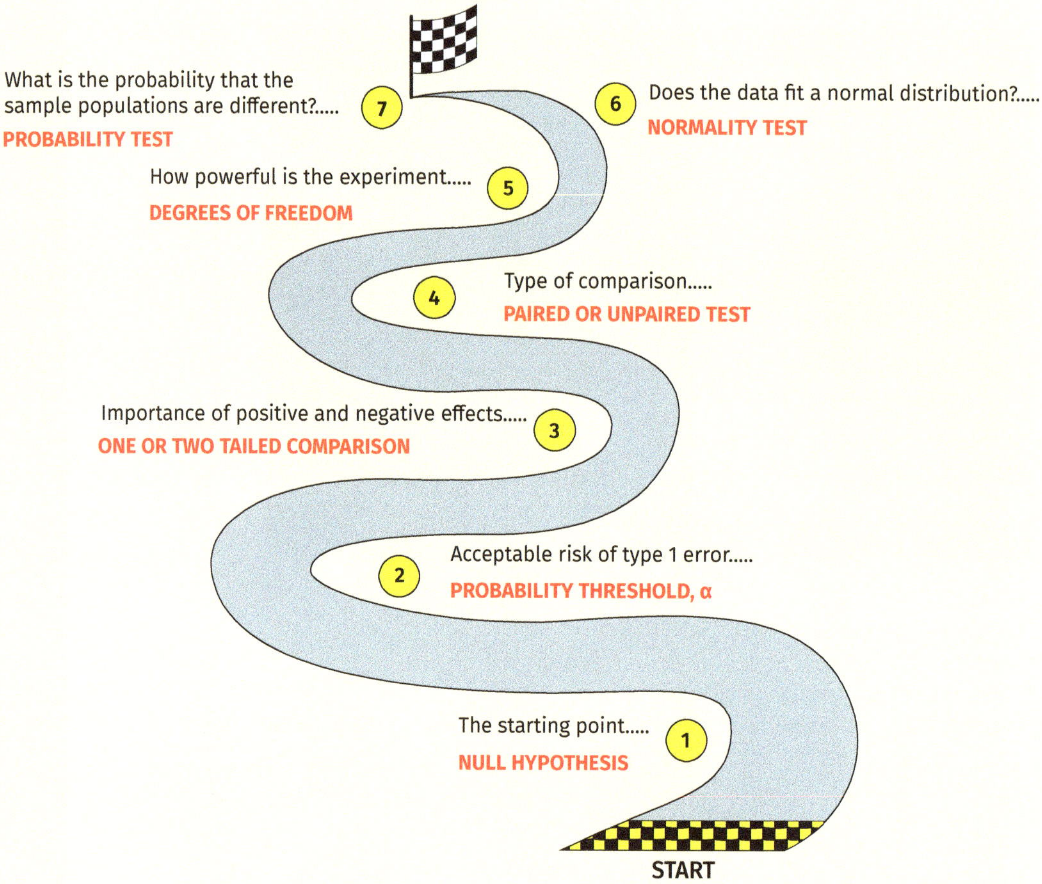

Checkpoints of hypothesis testing

Checkpoint 1: Plan your null hypothesis carefully

Science moves forward by testing whether null hypotheses (H_0) are retained (accepted) or falsified (rejected). In this way, H_0 works like a compass, giving the research path its direction. Remember that Popper's Empirical Falsification Theory (Chapter 1) tells us that different outcomes become possible as we discover new things and so H_0 is never 'true' or 'false' because these imply that the outcome is certain.

Pay close attention to the arrangement of terms in the null hypothesis because it marks the beginning of your journey and defines the route of discovery you will take. Let's think about a rewilding experiment in the Scottish Highlands which tests whether planting native trees like Scots pine and larch (reintroduction) helps to increase the population of Scottish wildcats (Figure 3.3). In this experiment, an area of replanted native trees (native) is compared to untouched mixed land-use habitat (mixed) and the number of wildcats recorded on a camera trap is measured in both areas over time.

Figure 3.3 Does rewilding improve habitat for Scottish wildcats? Statistics can help us find out—but we need the right null hypothesis!

A null hypothesis written as:

$$H_0 = \mu\, native = \mu\, mixed$$

predicts that any difference between the average number of wildcats in the two areas is due to chance alone.

This is much easier than stating:

$$H_0 = \mu\, native > \mu\, mixed$$

which says that mean wildcat numbers in the native area are greater than in the mixed area. Stated this way, there is no scope for observations being the same or lower in the native area.

The alternative, or research hypothesis (H_1), is a statement of the change that is predicted to happen if the null hypothesis is falsified. If, for example,

$$H_1 = \mu\, native \neq \mu\, mixed$$

we can conclude that difference between mean number of wildcats in both areas does not occur by chance. It is likely that manipulation of the independent variable, native tree habitat, produces real change in the dependent variable, wildcat population size. Notice how there is scope for wildcat numbers to be higher or lower in the native habitat compared to the mixed vegetation habitat. This is important statistically, because equal weight should be given to all possible outcomes. For all we know, planting native tree species might actually lower wildcat numbers.

Null hypotheses are set at the start of the experimental planning process and cannot be changed to suit the data after it has been collected or analysed. This would introduce special-cause variation (Chapter 2) which makes conclusions unreliable.

Checkpoint 2: Assigning a probability threshold

Confidence in an observation is expressed as a probability, or p-value, which can range between 0 and 1. As p-values approach 0 so the likelihood that an observation occurs by chance decreases, as does the risk of type I error.

How do we decide when to reject the null hypothesis? Scientists agree a threshold probability value which can be used to decide if a null hypothesis should be kept or rejected. This threshold is given the Greek symbol 'α' and is usually set at $p = 0.05$ following a convention set in 1925 by the famous statistician Ronald A Fisher who argued that:

> 'The value for which P = .05, or 1 in 20, is 1.96 or nearly 2; it is convenient to take this point as a limit in judging whether a deviation is to be considered significant or not. Deviations exceeding twice the standard deviation are thus formally regarded as significant.'
>
> Fisher, RA *Statistical Methods for Research Workers*, Oliver and Boyd, Edinburgh, 1925.

An α threshold value set at $p = 0.05$ tells us that we would expect to make a type I error (falsely retain the null hypothesis) on 1 of every 20 occasions that the experiment is conducted, or 5% of the time. If a p-value is below $\alpha = 0.05$ ($p < 0.05$) then we might reliably reject H_0. If $p > 0.05$ then the null hypothesis must be accepted because it remains the most probable explanation of the outcome.

Remember that p-values only provide a confidence test of the null hypothesis. It is incorrect to state that $p < 0.05$ suggests that the alternative hypothesis is true.

Checkpoint 3: One or two tails?

Does the study question allow for the possibility that outcomes could be either positive or negative (e.g. X and Y are different) or does it only look for effects in one direction (e.g. X is larger than Y)? If so, what might be missed by ignoring results in the other direction?

These questions are important because they determine how the risk of type I error is managed across the sample population (Figure 3.4).

To understand this, let's look at an important marine ecosystem on the Californian coast where the density of kelp forests, and the rich diversity of species living within them, depends on a balance between the sea urchins which graze the kelp, and sea otters which feed on the sea urchins.

A **two-tailed probability distribution** allows for the possibility of positive and negative outcomes. For example, a measure of Californian kelp forest density might produce data ranging from low density→no change→ high density depending on sea urchin grazing, the predation of the urchins by sea otters, and the shape of the seabed. The α threshold value is split equally between each end, or the two tails, of the normal distribution (Figure 3.4).

A **one-tailed probability distribution** assumes that change will always occur in the same direction. For example, a test of light exposure on kelp stem length is most likely to produce data ranging from no change → longer stems because kelp does not become shorter when exposed to light. Conversely, a study measuring seaweed weight on a rocky shore at low tide

Figure 3.4 One- and two-tailed tests allocate risk of type I error differently across the sample population (blue shading). Californian kelp forest density is affected negatively by sea urchin grazing and positively by sea otter predation of sea urchins and so has a two-tailed distribution of possible outcomes. Kelp frond response to light has a positive one-tailed outcome whereas changes in rocky shore seaweed mass whilst drying at low tide has a negative one-tailed outcome.

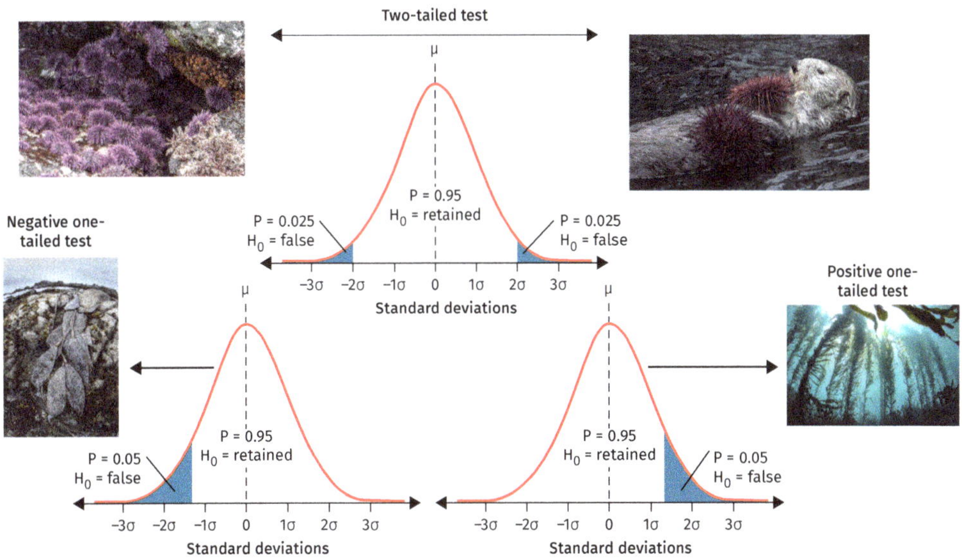

will produce a data range in the opposite direction because drying causes weight loss. Positive or negative one-tailed tests detect change more easily than two-tailed tests because the full distribution of the α threshold value sits entirely at one tail. For this reason, they also carry greater risk of type I error, particularly when sample sizes are small.

Statisticians follow rules to decide which test to use:

1. Decide at the start of the study if the data distribution is likely to be one- or two-tailed and explain your reasoning. The consequences of ignoring change in the other direction should be fully considered. If uncertain, a two-tailed test carries lower risk of type I error and is the default option.
2. A one-tailed test can never be used in place of a two-tailed test that had returned an insignificant result. Here, the investigator would be intentionally committing a type I error.

Checkpoint 4: How do I compare my groups?

A paired test looks for treatment effects in a matched sample population where the same individuals are compared before and after a treatment. This restricts variation to the treatment but carries the risk that background traits in the sample affect both data sets. Unpaired tests look for differences between two unrelated sample groups and so require a larger number of randomly selected samples to see any treatment effect above and beyond the natural variation of the two groups.

Deciding if a test is paired or unpaired becomes complicated for experiments using cell cultures and genetically modified animals or plants, because the definition of 'what is an individual' is debatable. For example, should an experiment using a cell line that has been split into two culture dishes be considered paired or unpaired? What about comparisons between genetically modified mice and their wild-type litter mates? Scientists often use social media forums like Researchgate to debate these questions openly.

Checkpoint 5: Degrees of freedom

You might imagine that numbers are free to vary in any direction in a study but this is not strictly true.

Here are the names of five fruit. Match the colour to the fruit:

1	Granny smith apple	Orange
2	Banana	Green
3	Tangerine	Periwinkle
4	Strawberry	Yellow
5	Akebia	Red

You might answer that granny smith apples are green, bananas are yellow, tangerines are orange and strawberries are red. Very few of us have ever heard of, or seen, an akebia; and what, exactly, is the colour periwinkle? By assigning answers to the other fruits, we are left with the inescapable conclusion that an akebia must be periwinkle in colour (it's bluey-purple). Of the five pairing options available, four can be matched in any combination but the remaining pair is constrained. In statistical language, we have five observations (n) with four **degrees of freedom**, or

$$\text{degrees of freedom (df)} = \text{n} - 1$$

Degrees of freedom tell us about the number of observations which can freely vary and, from this, the discriminating power of the experiment. Just like the fruit question, if all samples in a group are expressed as a percentage of the total, then all will be independent except one, which must bring the total to 100%. For a two-sample comparison, degrees of freedom are calculated for each treatment group:

$$\text{df} = n_1 + n_2 - 2$$

Checkpoint 6: Normality check

Parametric data have a distribution with shape or dimension. Think of the bell-shaped curve of the normal distribution where data fall symmetrically around the mean. Some types of data do not have a clear distribution pattern or might be skewed to one end of the range. Think of survey data, for example, where people are asked if they like something by voting yes or no. This type of data is called **non-parametric** because it does not follow a normal distribution and can vary widely.

Parametric statistics are used to look for differences between the means of two or more test groups if each population is normally distributed with a similar variance. Non-parametric statistics work by firstly assigning a rank to each data point so that the median of the ranks can be used as a measure of central tendency. It's a bit like strapping on a measuring tape. They then test to see if the median rank between the test groups is the same.

The normality check helps us to understand if our data should be analysed using parametric or non-parametric statistical methods. The Shapiro-Wilks test or the Kolmogorov-Smirnov test (see Chapter 4) are used to determine if the data fits a normal distribution. If not, a non-parametric test is required. This chapter gives examples of parametric and non-parametric tests which are commonly used in statistics.

Now that we have checkpoints for the route-map, how do we use them to test a null hypothesis? To understand this, it helps to look at how the method for comparing the means of two groups came about.

'Slàinte!' to Student's t-test

In 1886 the Guinness Brewery in Dublin was the largest in the world, with its famous stout consumed across Ireland and abroad. With this popularity came concerns about batch consistency, which was known to be very sensitive to variation in the brewing process. In a competitive market, brewers understood that an off batch would probably taste awful and could cost the company its customers and reputation.

The brewery owners recognized that a scientific approach was needed to understand the variables that defined the perfect glass of Guinness. In a stroke of management genius, they appointed William Sealy Gosset, a 23-year-old university graduate in mathematics and chemistry. Gosset understood that small-sized studies produce high sample standard deviations that do not accurately capture the population variance (see Chapter 2). To overcome this, he derived an equation for calculating probability distributions, known as **t-statistics**, from small sample sizes. In 1908 he published his discovery in a famous paper titled 'The Probable Error of a Mean' on condition that he concealed his name and his employer's identity so that other breweries would not know the Guinness Brewery's statistical secret for measuring batch quality. Gosset published under the pen-name 'Student' and the probability distribution of t-statistics he derived is the basis of **Student's t-test** for two sample comparisons (Figures 3.5 and 3.6).

Using Student's t-test to test a hypothesis

Perhaps to further protect the Guinness Brewery's industrial secret, Gosset's 1908 paper didn't mention barley production or brewing at all but instead used an example from a previously published study which tested whether right-handed (Dextro-, D) or left-handed (Laevo-, L) **stereoisomers** of a mild sedative called hyoscyamine hydrobromide produced a better night's sleep. Modern preparations of this compound are used in motion sickness

Figure 3.5 Gosset published his 1908 paper "The Probable Error of a Mean" under the pseudonym "Student"—not because the test was for students, but to hide his identity and so protect his method as an industrial secret for the Guinness Brewery!

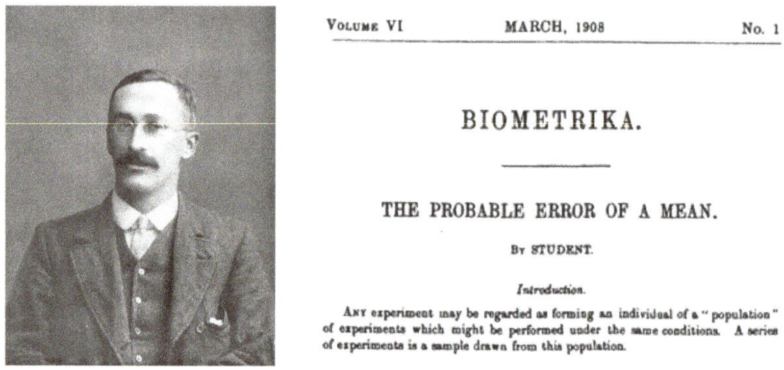

VOLUME VI MARCH, 1908 No. 1

BIOMETRIKA.

THE PROBABLE ERROR OF A MEAN.

BY STUDENT.

Introduction.

ANY experiment may be regarded as forming an individual of a "population" of experiments which might be performed under the same conditions. A series of experiments is a sample drawn from this population.

Figure 3.6 Student's t-distribution is used to calculate probability where sample sizes are small and population standard deviation is unknown. As sample sizes increase beyond 30, so the t-distribution begins to match the shape of the normal distribution. Student's t-test was developed to build consistency into the Guinness brewing method.

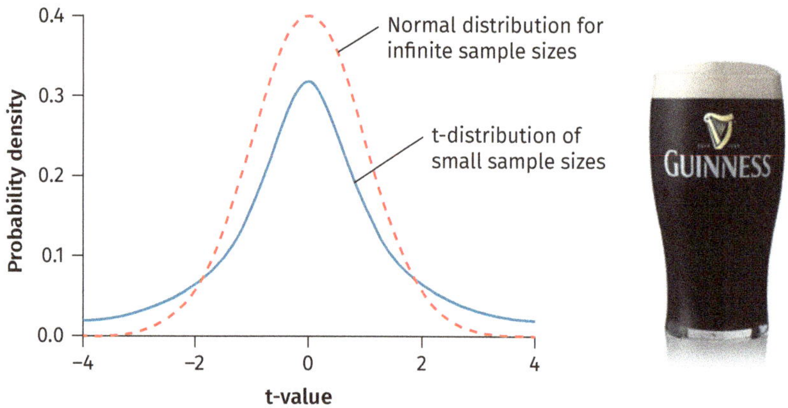

tablets. Let's work through his analysis of these data using the route-map checkpoints:

The route-map checkpoints are:

1. H_0: L and D stereoisomers of hyoscyamine hydrobromide produce the same amount of sleep in human subjects.
2. α: p = 0.05
3. Tails: Two-tailed because the null hypothesis allows that patients may either gain or lose sleep.
4. Paired or unpaired: Unpaired because different patients were used in the D and L treatment groups.

5. Degrees of freedom: Table 3.1 is from Gosset's paper and shows hours of sleep gained (positive numbers) or lost (negative numbers) for 10 patients after taking the same dose of each stereoisomer. Degrees of freedom for this experiment are:

$$\begin{aligned} df &= n_1 + n_2 - 2 \\ &= 10 + 10 - 2 \\ &= 18 \end{aligned}$$

6. Data distribution: Gosset did not check the distribution of these data, but a Shapiro–Wilks test confirms that it's normally distributed. Figure 3.7 shows each group as a box plot where the area of the box gives the upper and lower quartiles and the whiskers show the data range. Data points have been included to help you see how they fall around the sample mean ($\bar{x}$, horizontal red line) and median (horizontal black line).

When comparing two groups, the key measures are the distance between the sample means (signal), and the variation in the data (noise). You can think of the t-statistic as a signal-to-noise ratio, where values greater than 1 indicate that the signal is stronger than the noise. In Figure 3.7, the difference between the sample means for Dextro and Laevo isomers is the

Table 3.1 Additional hours' sleep gained by the use of hyoscyamine hydrobromide.

Patient	**Hyoscyamine Hydrobromide Stereoisomer**		**Difference**
	Dextro-	Laevo-	Laevo–Dextro
A	0.7	1.9	1.2
B	-1.6	0.8	2.4
C	-0.2	1.1	1.3
D	-1.2	0.1	1.3
E	-0.1	-0.1	0.0
F	3.4	4.4	1.0
G	3.7	5.5	1.8
H	0.8	1.6	0.8
I	0.0	4.6	4.6
J	2.0	3.4	1.4
Mean	0.75	2.33	1.58
SD	1.70	1.90	1.17

Gosset used data from a previously published study of sleep duration in human subjects given Dextro- or Laevo- hyoscyamine hydrobromide to show how his t-statistic works (data based on results from Gosset's original paper 'Probable error of a mean' in *Biometrika*, 1908).

Figure 3.7 Box plots and data distribution (open circles) for the hyoscyamine hydrobromide stereoisomer example used in Gosset's 1908 paper. The t-value calculation shown on the right takes the difference between the sample means and variance as a measure of the ratio of signal to noise.

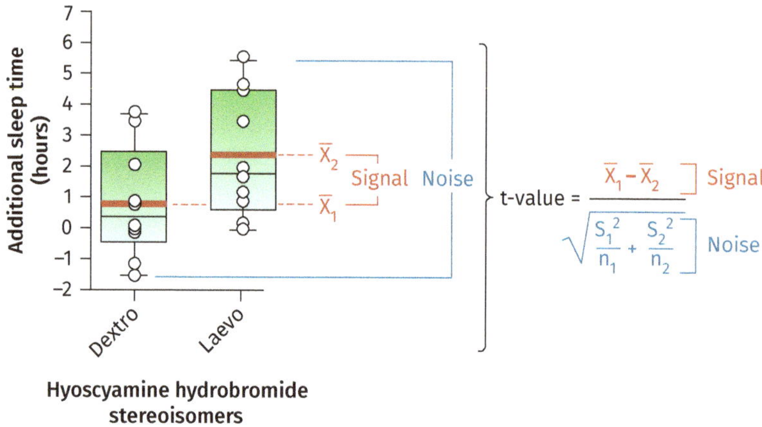

measurement 'signal' and is on the top half of the t-value equation, shown to the right of the plot. The variance is calculated from the sample standard deviation, s, of each data set (remember from Chapter 2 that the Greek symbol, σ, represents population standard deviation). By squaring each standard deviation and dividing by the number of observations for each sample, then taking the square root of the summed values for both data sets, we arrive at a measure of 'noise'.

We can use the data in Equation 3.1 to calculate the t-value for the experiment:

$$t - \text{value} = \frac{\bar{x}_1 - \bar{x}_1}{\sqrt{\frac{s_{1^2}}{n_1} + \frac{s_{2^2}}{n_2}}} \begin{matrix}]\text{Signal} \\ ]\text{Noise} \end{matrix}$$

$$=> \frac{0.75 - 2.33}{\sqrt{\frac{(1.7)^2}{10} + \frac{(1.9)^2}{10}}} \begin{matrix}]\text{Signal} \\ ]\text{Noise} \end{matrix}$$

$$=> \frac{-1.67}{\sqrt{0.81}} \begin{matrix}]\text{Signal} \\ ]\text{Noise} \end{matrix}$$

$t - \text{value} =>$ **−1.96**

The negative sign of the t-value shows the direction of the signal ($\bar{x}_1$ has a mean which is lower than $\bar{x}_2$) and since −1.96 is greater than −1, then there is more signal than noise (the mean difference is greater than the variance).

Table 3.2 gives the t-distribution as a table with degrees of freedom in the left-most column and α threshold probability values for one- and two-tailed tests across the top. Following our checkpoint rules, the calculated t-value of 1.96 is below the critical value of 2.101 needed to meet $\alpha = 0.05$ for 18 degrees of freedom and a two-tailed test, and so $p>0.05$.

Table 3.2 t-distribution table with critical values of t highlighted for 18 degrees of freedom.

	P value (α)			
One-tail	**0.1**	**0.05**	**0.025**	**0.01**
Two-tail	**0.2**	**0.1**	**0.05**	**0.02**
df				
1	3.078	6.314	12.710	31.820
2	1.886	2.920	4.303	6.965
3	1.638	2.353	3.182	4.541
:	:	:	:	:
16	1.337	1.746	2.120	2.573
17	1.333	1.740	2.110	2.567
18	1.330	1.734	2.101	2.552
19	1.328	1.729	2.093	2.539

We conclude that the null hypothesis should be retained: there is no statistical difference between the two groups.

How are statistical results reported?

Most scientists follow a convention for statistical reporting agreed by the American Psychological Association, or APA, which states that t-test results should be given as: test type (degrees of freedom) = the t-statistic, p = p-value.

Following this format, the conclusion from Gosset's t-test example is:

> There was no significant effect for sleep, t(18) = –1.96, $0.05<p>0.1$ between subjects receiving the Dextro ($\bar{x}$ = 0.75, s = 1.7) and Laevo ($\bar{x}$ = 2.33, s = 1.9) stereoisomers of hyoscyamine hydrobromide.

Why are non-parametric tests important in biology?

Parameters define the boundaries of a system. In migrating birds, for example, body fat, time, and distance are three parameters which determine how far a bird can fly before it must stop to feed. We can describe fuel consumption in birds as a parametric measure of migration behaviour.

Many things in biology, however, cannot be studied using defined parameters like these because either:

- they do not have a measurable size or boundary (dimension);
- they do not align to an arithmetic number scale; or
- they lack a normal distribution.

Furthermore, life processes are often controlled by things which are not fully understood, and so we may not know how they should be measured. Non-parametric tests allow us to side-step assumptions about study size and data distribution by placing emphasis on the ranked order of the data in the study groups. Modified probability distribution tables are then used to compare the calculated statistic to a selected critical value in a similar way to the t-test. Let's look at how an imagined follow-up experiment to Case Study 3.1 might be investigated using a non-parametric method.

Case study 3.1

Polyandry protects *Petrolisthes*? The statistical road map in action

Petrolisthes, the porcelain anemone crab, lives in the intertidal zone of many world oceans (Figure A) and, like most crustaceans, it breeds by external fertilization. This means that egg broods are sometimes fertilized by more than one male, so increasing the genetic diversity of the next generation. This breeding behaviour, known as polyandry (many fathers), is often described as 'bet-hedging' because it improves the species tolerance to heat, pH, desiccation, and oxygen stress which occur daily in the intertidal zone.

Researchers from San Francisco State University asked whether polyandry improves the ability of the flat porcelain crab, *P. cinctipes*, to survive acute intertidal environmental stress. Their research hypothesis was:

Figure A A porcelain anemone crab.

'H_1: multiple paternity is under strong positive selection in *P. cinctipes*, aiding in bet-hedging broods to withstand the abiotic stresses of the intertidal zone.'

The research approach to the first experiment in the study is shown in Figure B. The study uses temperature increase as an easily controlled independent variable, typical of the stress that occurs on a rocky shore between tides. The statistical checkpoints are declared in the study methods:

Checkpoint 1: The null hypothesis: The study begins by asking if heat stress alters embryo survival in broods that have been fertilized by one male

Figure B Paired experimental design for testing the effect of heat stress on embryo hatching success for single paternity (SP) and multiple paternity (MP) broods. Life-cycle stages of the porcelain crab are shown on the left and the results of the experiment are shown in the box plot below. The asterisk indicates that there is a statistically significant effect between the SP broods. Source: J. Stilman/SUSF, USA/2, Yochachonis et al., *J. Crust. Biol.* 40(6): 684–691, 2020

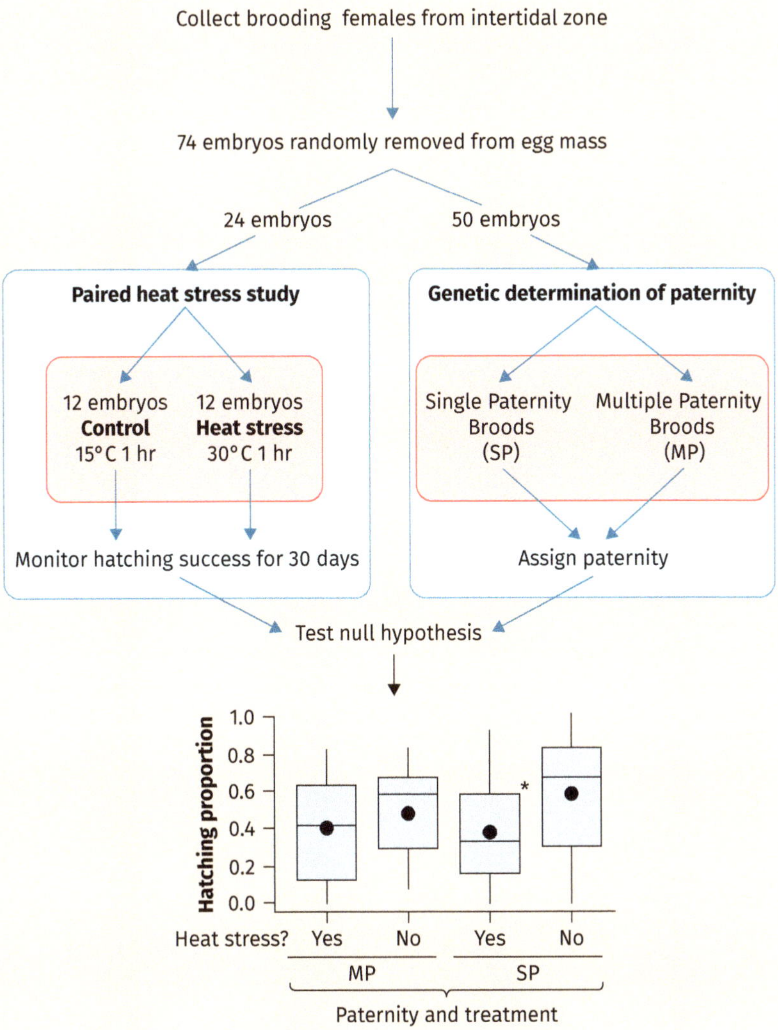

(single paternity, SP) or several males (multiple paternity, MP). Hatching success was measured as the number of larvae hatched divided by the number of eggs produced by a single crab and served as a measure of breeding success. The null hypothesis is:

H_0: 'heat shock does not reduce hatching'.

Checkpoint 2: Risk of type I error (α): Set before the experiment at $p \le 0.05$.

Checkpoint 3: One-tailed: The study assumes that heat stress has either no effect or decreases hatching success in SP and MP broods. The opposite outcome, that heat stress might increase hatching success, is not considered because earlier research showed that excessive heat is lethal because proteins become denatured.

Checkpoint 4: Paired experiment: the dependent variable, hatching success, was tested against the independent variable, heat stress, in embryos from the same brood. The paired design of the experiment is shown in Figure B. For each brood female, 24 embryos were removed and split into two groups of 12. A control group was maintained at 15 °C for 1 hr and a test group was maintained at 30 °C for one hour. Embryos were then maintained in separate dishes for several weeks to measure hatching success. Other embryos from the same brood were used to establish single or multiple paternity by genetic analysis.

Checkpoint 5: Degrees of freedom: Student's t-test assumes that data for each group are normally distributed with a similar variance; however, the hypothesized genetic diversity between the SP and MP groups means that it is unsafe to assume that the variance is the same. For this reason, the study uses a variation on Student's t-test known as Welch's t- test which assumes unequal group variance and calculates degrees of freedom using the standard deviations between the groups rather than number of observations. Degrees of freedom reported for this study are 15 using the Welch's t-test calculation.

Checkpoint 6: Normality test: Previous studies showed that embryo mortality data is normally distributed in both groups. A parametric test is appropriate.

Welch's t-test was used to determine if the null hypothesis should be kept or rejected following heat stress applied to each group.

The box plot in Figure B gives measures of central tendency for each treatment (the dot shows the mean, horizontal bar is the median) and the quartile ranges (box height is the interquartile range and the whiskers show the overall range for all four quartiles). The authors describe their findings APA style which is the way statistical results should be reported in academic writing:

> 'The mean effect of heat stress on the number of embryos hatching was twice as large in broods with single paternity (paired one-tailed t-test, $t(15\ df) = -2.97$, $p = 0.005$; mean difference = −2.44) than multiple paternity (paired one-tailed t-test, $t(10\ df) = -1.54$, $p = 0.077$; mean difference = −1.09).'
>
> Yochachonis et al., *J. Crust. Biol.* 40(6): 684–691, 2020

It shows that H_0 must be accepted for the multiple paternity group comparison because the effect of heat stress on hatching success is no different to the mean difference that would occur by chance. H_0 is rejected, however, for the single paternity group comparison, suggesting that hatching success in this group is sensitive to heat stress. We conclude that single paternity embryo hatching success is reduced by heat stress but multiple paternity hatching success is unaffected.

Where next on the route-map? The study uses this information to test a sequence of increasingly detailed research questions, shown in Figure C.

Figure C Experimental road map based on a study by Yochachonis et al., *J. Crust Biol.* 40(6), 684–691, 2020 to determine if polyandry is an advantage in intertidal porcelain crabs facing heat stress.

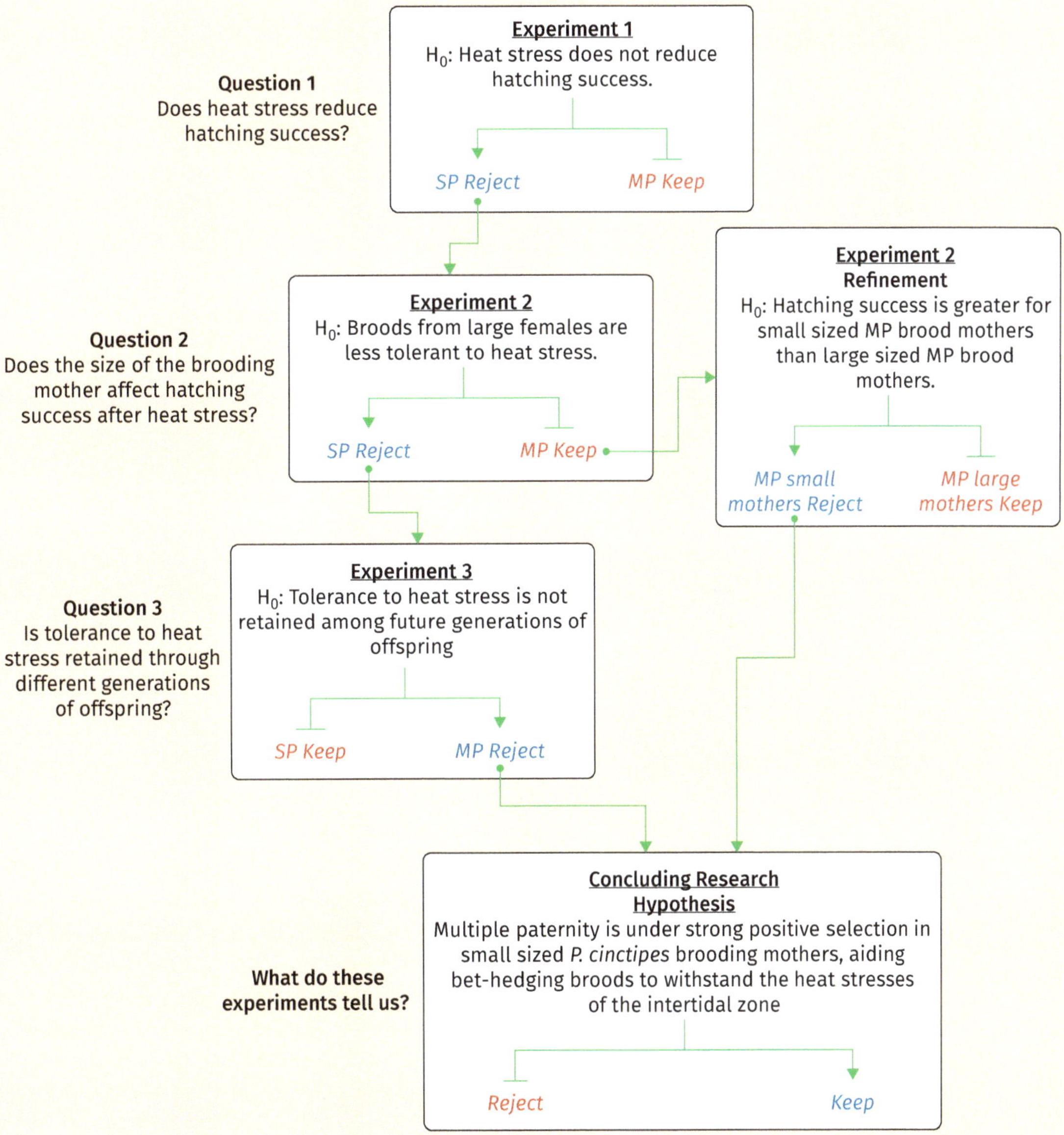

Notice how null hypotheses drive the questioning process towards a key experiment which explores how tolerance to heat stress is conserved in generations of offspring from multiple paternity broods which have never been exposed to high heat conditions. These experiments support the research hypothesis that:

> 'Multiple paternity is under strong positive selection in small-sized *P. cinctipes* brooding mothers, aiding bet-hedging broods to withstand the heat stresses of the intertidal zone.'

The path is open for more research to explore if the same principles apply to other stressors and to find the molecular mechanisms which are behind the survival effect.

Pause for thought

- Do you think bet-hedging in other species that reproduce by external fertilization, like corals, frogs, or fish, might help them to adapt quickly to climate change? How might you test your idea?

How do we assign statistical significance to non-parametric information?

The research example in Case Study 3.1 concludes that a multi-paternity (MP) genetic lineage favours tolerance of heat stress whereas single paternity (SP) does not. Heat shock protein-70 (HSP70) is a **molecular chaperone** which stabilizes cell proteins during heat stress and helps to clear away damaged proteins after it has passed. Its appearance in the cell is normally induced by heat shock, but under some circumstances it is continuously present, or constitutive, providing constant protection to cell proteins. Might constitutive HSP70 explain the MP brood tolerance to heat stress which runs through generations?

We can test this using the following hypotheses:

H_0: HSP70 cannot be detected in SP and MP broods five generations after a heat shock stress.

H_1: HSP70 is constitutively expressed (always present) only in MP broods five generations after a heat shock stress.

In our imagined experiment, five SP and five MP broods were carried through from a heat shock exposure at generation 0, to the end of the experiment at generation 5. Single paternity was maintained for both groups after the heat shock. Within each brood, up to 10 embryos were selected at random and analysed for the presence or absence of the protective protein HSP70. **Western blotting**, a technique used to detect specific proteins, is used to identify HSP70. If present in the embryo, HSP70 appears as a distinct band on the blot. The number of positive bands for each brood were counted and the results are shown in the table in Figure 3.8.

Figure 3.8 An imagined follow-up experiment from Case Study 3.1 to determine if HSP70 expression is still present (constitutive) five generations after heat shock in embryos from multiple (MP) and single paternity (SP) broods. Western blotting is used to detect HSP70 expression, and the table and frequency histogram reports the number of HSP70 positive embryos.

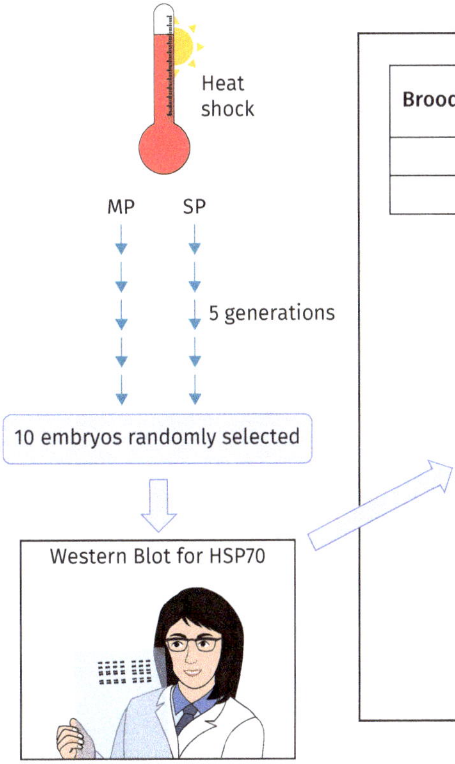

Brood Paternity	Number of HSP70 positive embryos				
	Brood #1	Brood #2	Brood #3	Brood #4	Brood #5
SP	0	1	4	1	2
MP	4	7	5	2	10

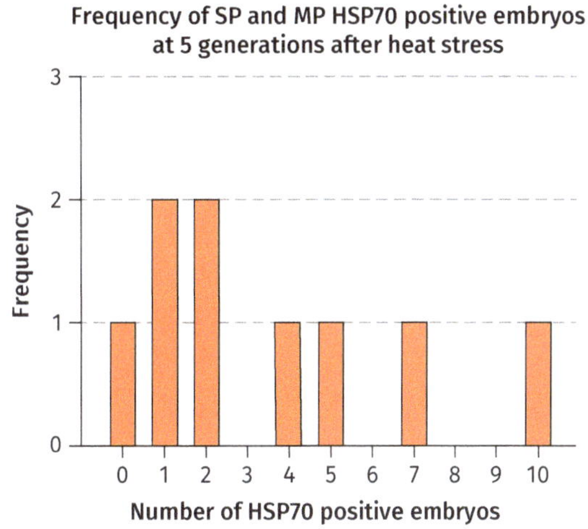

The frequency histogram shows that the data do not follow a normal distribution and that the sample size of five in each paternity group is small.

So how do we test the research hypothesis if the data are non-parametric and so can't be measured against a natural scale?

Assigning significance to ranked non-parametric data using the Mann–Whitney U Test

The **Mann–Whitney U Test** is a non-parametric method used to determine if two unpaired groups are likely to come from the same population. The aim is to calculate a critical test statistic, known as U, using the ranked order of the data for each test group. A table of critical values of U is used to determine the probability threshold for rejecting H_0, where

H_0 = the medians of the two populations are equal.

The test combines the data from both groups into a single ranked sample while still keeping track of which observations came from each group. Data are then ranked from low to high.

Table 3.3 Arrangement of all observations from both groups into a single, sequentially ordered data set.

	Brood Paternity	Number of HSP70 positive embryos									
		Brood #1	Brood #2	Brood #3	Brood #4	Brood #5					
Original Data	SP	0	1	4	1	2					
	MP	4	7	5	2	10					
Data Sequence	SP	0	1	1	2		4				
	MP					2		4	5	7	10

Using the porcelain crab brood paternity data from Table 3.3 the pooling and ranking procedure works like this:

Step 1: Organize the data from both groups into a sequence from the smallest value to the greatest, indicating clearly which group each data point comes from:

Step 2: Rank the data giving the lowest data point a ranked value of 1, next in the sequence a value of 2, and so on. Where the same values occur more than once, the average of the assigned rank score is taken. For example, the data sequence of the two groups is:

0 - 1 - 1 - 2 - 2 - 4 - 4 - 5 - 7 - 10

Then the ordered sequence would be:

1 - 2 * -3 * -4 ** - 5 ** - 6 *** -7 *** -8 - 9 - 10

The stars show where the same data points occur in the sequence. To assign a rank to these values, take the average of their positions in the number sequence. For example, the duplicate '1' values in positions 2 and 3 would be ranked as: 2+3/2 = 2.5. The same idea applies to any number which occurs multiple times in the sequence.

So, in our example:

From the original sequence

$$0\,(\text{lowest data point}) = \mathbf{1}$$

1 = the numbers in positions 2 and 3 are the same so the ranked numbers are

$$(2+3)/2 = 5/2 = \mathbf{2.5}$$

2 = the numbers in positions 4 and 5 are the same so the ranked numbers are

$$(4+5)/2 = 9/2 = \mathbf{4.5}$$

4 = the numbers in positions 6 and 7 are the same so the ranked numbers are

$$(6+7)/2 = 13/2 = \mathbf{6.5}$$

$$5 = \mathbf{8}$$

$$7 = \mathbf{9}$$

$$10 = \mathbf{10}$$

The final rank order of the sequence is shown in Table 3.4.

Note that we have five observations in each group (n_{SP} and $n_{MP} = 5$) and the value, R, is the sum of the ranks for each group. Having ranked the data,

Table 3.4 Ranking the combined data set after arranging all observations in ascending order.

	Brood Paternity	Number of HSP70 positive embryos: Brood #1	Brood #2	Brood #3	Brood #4	Brood #5						
Original Data	SP	0	1	4	1	2						
	MP	4	7	5	2	10						
Data Sequence	SP	0	1	1	2		4					
	MP					2		4	5	7	10	
Ranked Data	SP	1	2.5	2.5	4.5		6.5					Σ ⇒ R_{SP} = 17
	MP					4.5		6.5	8	9	10	Σ ⇒ R_{MP} = 38

we are now ready to use the Mann-Whitney U Test to find out if the two groups are from the same or different populations.

The U statistics for the SP and MP groups are calculated like this:

Using the final ranking sequence and R values, values for U are:

For the single paternity group, see Equation 3.2:

$$U_{SP} = n_{sp}n_{mp} + \frac{n_{sp}(n_{sp}+1)}{2} - R_{SP}$$

$$U_{SP} = 5(5) + \frac{5(6)}{2} - 17 = 23$$

For the multiple paternity group, see Equation 3.3:

$$U_{MP} = n_{sp}n_{mp} + \frac{n_{mp}(n_{mp}+1)}{2} - R_{MP}$$

$$U_{MP} = 5(5) + \frac{5(6)}{2} - 38 = 2$$

U values range upwards from 0, where 0 indicates complete separation between the groups and higher values trending towards increasing similarity. The smaller of the two U values is taken as the test statistic, called the critical value of U or U_{crit}. This can be used to test if the null hypothesis should be retained or rejected.

In our porcelain crab HSP70 data, $U_{crit} = 2$, but can we reject the null hypothesis? Following the statistical checkpoints, we assume a two-sided level of significance, $\alpha = 0.05$ for a sample size of $n = 5$ for each group.

Table 3.5 shows that the critical value for $P \leq 0.05 = 2$ and so we reject H_0 for the multiple paternity group.

Conclusion: HSP70 expression in the multiple paternity group follows a different population distribution to the single paternity group. There are reasonable grounds to set a new research hypothesis to determine if the presence of HSP70 extends heat tolerance and survival in multiple paternity porcelain crab broods.

Applying non-parametric tests to paired data

Non-parametric tests can be used to compare groups where the data is paired but does not follow a normal distribution. Two common non-parametric

Table 3.5 To find the critical values of U for α= 0.05 (two-tailed), find the value at the intersect between the group 1 sample size (n_1) and group 2 sample size (n_2).

		n_2						
		2	**3**	**4**	**5**	**6**	**7**	**8**
n_1	2							0
	3				0	1	1	2
	4			0	1	2	3	4
	5		0	1	2	3	5	6
	6		1	2	3	5	6	8
	7		1	3	5	6	8	10
	8	0	2	4	6	8	10	13

methods for comparing paired groups are the Sign Test and the Wilcoxon Signed Rank Test. Let's consider a paired study where data is skewed in the measurement groups.

In areas like the Scottish Highlands, alpine snow patches provide controlled climate conditions and essential nutrients for micro-invertebrates and plant and lichen communities. However, they are sensitive to erosion, unpredictable weather patterns, and human impact. The National Vegetation Classification (NVC) identifies two species of moss, the Northern Haircap (*Polytrichum sexangulare*) and Starke's Fork Moss (*Kiaeria starkei*) as unique to long-lasting snow beds. Scientists study the distribution of these specialist mosses to monitor the condition of snow bed habitats (Figure 3.9).

In our study, the average ground cover area for each snow bed was gathered using the random quadrat method (B in Figure 3.9). The data table, Table 3.6, and box plot, Figure 3.10, shows average ground cover for Northern Haircap moss, measured in May of two consecutive years for 10 different late season snow beds. Notice how, in Figure 3.10, the mean (dotted line) and the median (solid line) are separated in the box plots due to skew in the data set. Yet both measures of central tendency look similar between the two study years despite obvious differences in the distribution of the data, shown by the interquartile range.

The Sign Test

This test by-passes problems of data distribution by simply testing if a change of any size has happened. It does this by finding the difference between two events and then adding up the number of positive and negative numbers. If the n value and the number of positive and negative outcomes are known, a binomial distribution table can be used to calculate the probability of success (positive outcomes, p) or failure (negative outcomes, $q = 1 - p$).

Figure 3.9 Late season snow fields such as this example in the Scottish Highlands support a unique snowbed habitat defined by two species of moss, the Northern Haircap and Starke's Fork Moss. Snowbeds are often studied using transects from the edge of the snowbed (A) or randomly placed 1m^2 quadrats in a defined radius of the snow margin (B).

Table 3.6 shows results from a sampling experiment which measured Northern Haircap mat areas for 10 different year-round snow beds. The measurements were taken in the spring of two consecutive years using quadrats placed randomly around each snow bed. Let's use the Sign Test to find out if Northern Haircap mat area has changed significantly between year 1 and year 2.

Step 1:

What are our hypotheses? Null hypothesis (H_0): there is no difference in the moss mat area in year 1 compared to year 2.

The histogram shows that the mean value and interquartile range are larger for year 1 compared to year 2 even though the median values appear

Table 3.6 and Figure 3.10 Table and bar graphs showing average Northern Haircap mat area for 10 different late season snow beds measured in May of two consecutive years. The box plot shows the distribution of the data for each group where the horizontal solid line is the median, dotted line the mean, box margins the inter-quartile range, and the whiskers show the data range. Outliers are shown as single data points above the whiskers.

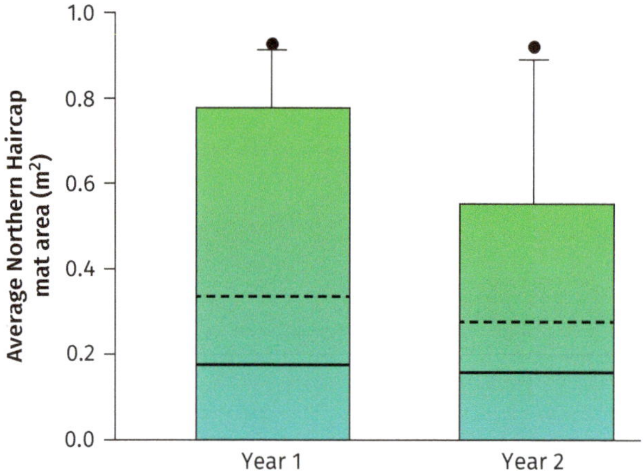

very similar. Might there be a difference between the years? We can test this as follows:

Research hypothesis (H_1): moss mat area is greater in year 1 compared to year 2.

Step 2:

Table 3.6 shows the steps for calculating the direction of change in the data. For each snow bed, subtract the mat area value of Year 1 from Year 2

Table 3.6 Average Northern Haircap mat area (m^2).

Snow bed number	Year 1	Year 2	Difference (Year 1 - Year 2)	Change
1	0.76	0.63	0.13	+
2	0.83	0.92	-0.09	-
3	0.42	0.25	0.17	+
4	0.20	0.31	-0.11	-
5	0.00	0.05	-0.05	-
6	0.15	0.00	0.15	+
7	0.00	0.00	0.00	nd
8	0.92	0.53	0.39	+
9	0.05	0.06	-0.01	-
10	0.01	0.00	0.01	+

to obtain a difference. Record the gain or loss of mat area as a positive or negative sign in the 'Change' column. If no change has occurred, record nd (no difference).

Step 3:

Work out your sample size (n) by adding up the number of observations and subtract any which showed no difference. From Table 3.6, there are 6 positive signs, 3 negative signs, and 1 nd and so n = (5 + 4 +1)–1 = 9. Notice that we have a mix of positive and negative numbers and so our test is two-sided. A one-sided test would be used if all the number differences fell in one direction.

Step 4:

To find the p value, you will need a binomial distribution table, and a link to one is provided in the further reading at the end of this chapter. The tables look complicated but all they really report is sample size (n), the number of positive observations (x), and the probability of success for the sample size (P). You can use an online calculator but we will use a table in this example to show you how the calculation works.

Binomial distribution tables list the probabilities associated with the null hypothesis *across the top row*. Our null hypothesis states that there is no difference between the study groups so in this case we would expect to encounter either a positive or negative value half the time, or p = 0.5. We will use this column in the binomial distribution table as our reference to find the critical p value for our data.

Step 5:

Our research hypothesis says that moss mat area will be larger in year 1 than year 2 and so we can use the table to find out if we are more likely to encounter a positive result compared to the null hypothesis. From Table 3.6, there are 5 positive outcomes and the sample size is n = 9 so we can apply these numbers to the binomial distribution table, an extract of which is given in Table 3.7.

Table 3.7 Extract of a binomial distribution table showing how many positive observations would be needed for P = 0.5.

Sample Size	Number of positive observations	P = 0.5
n = 9	0	0.002
	1	0.020
	2	0.090
	3	0.254
	4	0.500
	5	0.746
	6	0.910
	7	0.981
	8	0.998
	9	1.000

Conclusion: The probability of encountering the null hypothesis is 0.746 which is well above the critical $P_\alpha = 0.05$ value we would need to meet for accepting statistical significance. We must accept the null hypothesis because there is no qualitative difference in moss distribution around the snow beds between study year 1 and 2.

The Wilcoxon Signed Rank Test

This follows the same idea as the Sign Test but with the advantage that it takes account of the size of the difference between the two test groups. Here, the sum of the positive (W+) and negative (W−) ranks is used to calculate a value of W which is compared with a critical value. The following steps take you through the method for ranking and assigning probability to the results which are collected together in Table 3.8.

Step 1:

Use the differences in moss mat cover recorded in years 1 and 2 to organize the data from the smallest to the largest difference (it doesn't matter if it is a positive or negative difference at this stage). The table shows that snow bed 7 has the smallest difference (0) and snow bed 8 the largest (0.39).

Step 2:

Now assign a rank order number from lowest to highest. Snow bed 7 showed no change in moss mat cover so this is not assigned a rank and is labelled 'nd' (no difference). Snow beds 9 and 10 are next in order but both have the same difference value of 0.01 if we ignore the direction of change (we will deal with that later). If we ranked snow bed 9 as 1 and snow bed 10 as 2 their contribution to the Wilcoxon Test would not reflect their equal values.

Table 3.8 Wilcoxon Signed Rank Test (W).

<table>
<tr><th rowspan="2">Snow bed number</th><th rowspan="2">Difference (year 1-year 2)</th><th rowspan="2">Rank</th><th rowspan="2">Rank with sign</th><th>W+</th><th>W-</th></tr>
<tr><th>Sum of positive ranks</th><th>Sum of negative ranks</th></tr>
<tr><td>7</td><td>0</td><td>nd</td><td>nd</td><td rowspan="10">31.5</td><td rowspan="10">-13.5</td></tr>
<tr><td>9</td><td>-0.01</td><td>1.5</td><td>-1.5</td></tr>
<tr><td>10</td><td>0.01</td><td>1.5</td><td>1.5</td></tr>
<tr><td>5</td><td>-0.05</td><td>3</td><td>-3</td></tr>
<tr><td>2</td><td>-0.09</td><td>4</td><td>-4</td></tr>
<tr><td>4</td><td>-0.11</td><td>5</td><td>-5</td></tr>
<tr><td>1</td><td>0.13</td><td>6</td><td>6</td></tr>
<tr><td>6</td><td>0.15</td><td>7</td><td>7</td></tr>
<tr><td>3</td><td>0.17</td><td>8</td><td>8</td></tr>
<tr><td>8</td><td>0.39</td><td>9</td><td>9</td></tr>
</table>

Table 3.9 Alpha value.

n	0.01	0.025	0.05
5	-	-	-
6	-	-	0
7	-	0	2
8	0	2	3
9	1	3	5
10	3	5	8

The solution is to assign the average of their ranking sequence to both, or (1+2)/2 = 1.5. Always treat equal values the same way by averaging their rank in the sequence.

Step 3:

In the next column, assign a positive or negative sign to the rank depending on the direction of change indicated by the year 1–year 2 difference.

Step 4:

Calculate W+ as the sum of the positive ranks and W− as the sum of the negative ranks. The null hypothesis predicts that W+ = W− and our calculated values are W+ = 31.5 and W− = −13.5. As with the other two comparison tests, a table of critical W values is used to determine if the smaller of the two calculated values of W support or retain the null hypothesis.

Step 5:

Critical values for the Wilcoxon Signed Rank Test are given in tables and a link is provided to a suitable source in the further reading at the end of this chapter. To determine the critical value of W, we need the sample size (n = 9, omitting the sample with no difference, nd), the significance threshold, $\alpha = 0.05$ and a two-sided comparison. An excerpt of the W critical values table is shown in Table 3.9 and returns the number 5 for our test. The H_0 should be rejected if $W \leq 5$. Our lowest calculated value of W is −13.5 and so the null hypothesis must be retained.

Conclusion: Even considering the size of the differences there is no quantitative difference in moss distribution around the snow beds between study year 1 and 2.

Biostatistical comparisons for the Omics era

The complete sequencing of the human genome by the Human Genome Project stands as one of the landmark scientific achievements of the modern era. Over six years, researchers accurately read all 3 billion DNA bases of the human genome. During this period, DNA and RNA sequencing technologies advanced dramatically, allowing scientists to read long stretches of nucleotides in minutes rather than the weeks such work once required. It has helped us identify 3.7 million human single nucleotide polymorphisms, or SNPs, which are scanned in genome-wide association studies (GWAS) to find mutation patterns that associate with disease and help develop personalized treatments.

Perhaps most importantly, however, the Human Genome Project has advanced understanding of biomolecular interactions through the study of 'Omics', the collective term for the study of DNA (genomics), RNA (transcripomics), proteins (proteomics), kinases (kinomics), and metabolism (metabolomics). You can find out more about many of the 'Omics' in Genomics, another title in this primer series.

Figure 3.11 Network parameters which feed statistical models of the interactome. A complete 'global' protein interactome containing 13,460 proteins and 141,296 interactions is shown in the lower part of the figure and works like a map to show how cell processes change in health and disease.

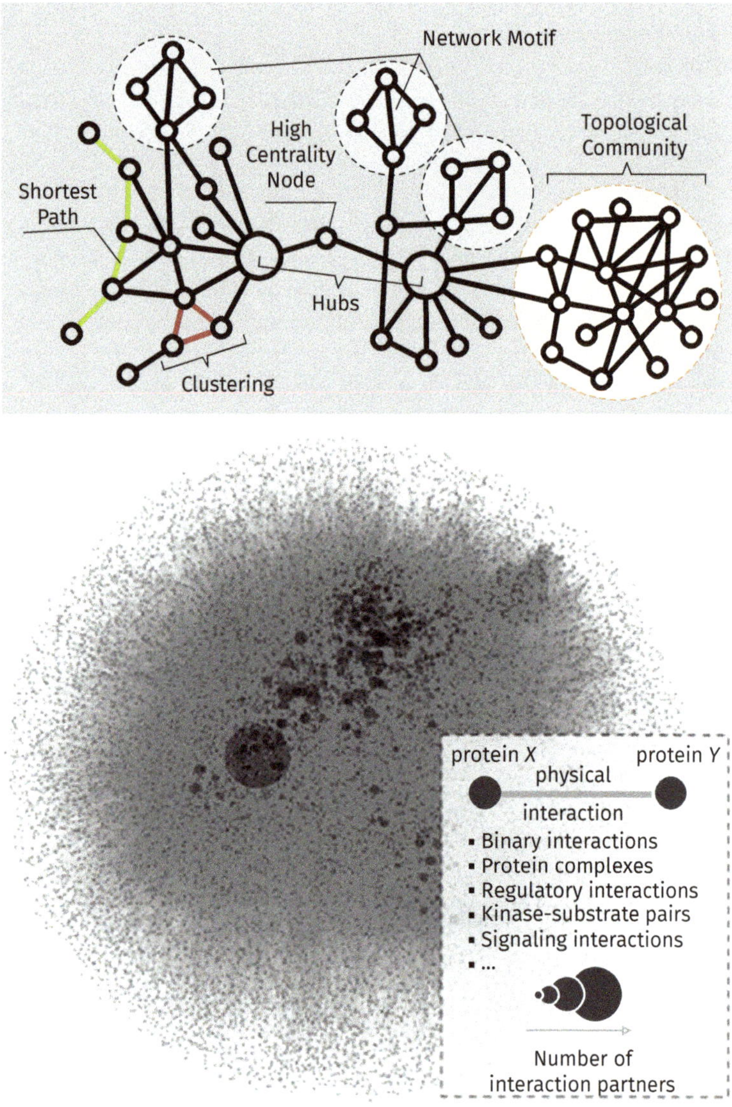

The **interactome** encompasses all of these fields to explain how biomolecules communicate in the whole cell. The analysis of these relationships, often using non-parametric methods, has led to a new age of biostatistical network modelling which has revolutionized our understanding of disease and its treatment.

Network modelling begins with the collection of information about how genes and proteins interact from published scientific studies and database repositories. From here, we can describe the different types of interactions that make up the interactome (Figure 3.11). A **node** is an interacting point between biomolecules (e.g. proteins) which has several links attached to it, known as **degrees**. The most energetically favoured connection between nodes is known as the **shortest path**. Nodes which connect different parts of the interactome are known as **hubs** and measures of their centrality in the interactome describe their structural importance in connecting parts of the network together. **Clustering** describes the tendency for parts of a network to interact with each other, often involving a triangle of nodes. **Network motifs** are repeated interactions which often define the unique characteristics of a part of the interactome. **Topological communities** are groups of nodes that have more connections between themselves than to the rest of the network.

Non-parametric statistics are particularly helpful for scoring interactions and assessing their importance in the interactome.

Bigger picture 3.1

Biostatistics and the new world of network medicine

Rare genetic diseases affect fewer than 1 in 1500 people and are often caused by single nucleotide omissions or substitutions that change the expression of a gene and its downstream protein product. They are often described as **phenotypically pleiotropic** because the change in protein behaviour can have multiple effects on cells and body systems. For example, a single amino acid substitution in the CFTR ion channel which causes cystic fibrosis disease doesn't just inhibit fluid secretion in organs like the lung and the gut, but also affects many other processes like inflammation, pathogen defence, growth, nutrient uptake, and metabolism. For the same reasons, it is often found that many rare diseases share similar medical characteristics despite having very different genetic origins.

A group of researchers from the USA used an omics approach to explore how distinct rare diseases might involve proteins from overlapping regions, or neighbourhoods, of the interactome. They argued that if two different diseases are affected by the same protein neighbourhood then changes caused by one disease will affect protein interactions in the other, giving both diseases shared medical characteristics.

Figure A Diseases can be mapped to the interactome to show how protein communications are shared. Common cell interactions are involved where disease modules overlap. The expanded figure shows how the immunoglobulin A pathway explains common characteristics between asthma and celiac disease.

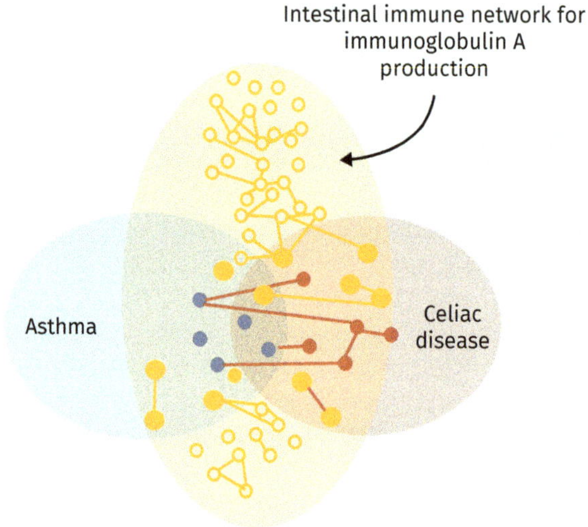

They used a combination of network modelling and non-parametric statistics such as the Mann–Whitney U Test to explore how much biological similarity was present for different biological processes, molecular functions, shared cell components, protein profiles, symptom similarity, and **comorbidity** (presence of more than one disease in a person at a time). Figure A shows how the information comes together to reveal similarities between asthma and coeliac (or celiac) disease, which share common protein neighbourhoods of the interactome. The yellow nodes of the intestinal immune network show how the immunoglobulin A production pathway plays a biological role in both diseases. Immunoglobulin A is an important molecule in the immune response—large amounts are found on the surface of the gut lining, where it inactivates many pathogens before they get into the cells of the body—but it is also sometimes involved in auto-immune conditions. Advances in network statistics allowing greater interpretation of the interactome underpin the newly emerging discipline of network medicine allowing deeper insight into genome-wide association studies, drug targeting, and use of old drugs for new purposes.

Pause for thought

- How do you think that the parametric and non-parametric statistical ideas in this chapter might be used to describe reactions between biomolecules in an interactome?

Chapter summary

- Hypothesis testing provides a structured framework for making experimental decisions. In practice, it acts like a road map, guiding researchers through the process of discovery.
- The null hypothesis is the starting point on this journey. It must be defined accurately to allow for all possible alternative outcomes.
- A t-value works like a signal-to-noise ratio where the signal is the distance between two normally distributed sample means and the noise is the variance of each sample.
- Student's t-test gives us a way to assign probability to the difference between the means of two normally distributed populations.
- In biology, parametric data can be measured against a scale, shape, or dimension of some kind but non-parametric data has no dimensions.
- Non-parametric tests use the ranked order of observation to assign statistical significance.
- Parametric and non-parametric tests provide the foundation for studying the new era of biological interactions known as Omics.

Discussion questions

Statistical tests are a tool for discovery. They help you to make informed decisions about unknown things using the evidence available to you. In short, they are empowering. Here are three tasks to encourage you to think about the ways in which you might use statistics to explore your own ideas in biology.

1. Use the six-step route-map to design an unpaired experiment which could be tested using Student's t-test. For example, you might want to find out if caffeine in fizzy drinks can improve problem-solving ability. Think how you might use caffeinated and caffeine-free versions of the same fizzy drink with a cognitive test that has an easily measured data output. Timed completion of a Stroop test where the subject states colours of words in a printed list is one example of a simple and effective cognitive test which you can make yourself. When using Student's t-test to determine the t-value and statistical significance, think about the assumptions of your experiment, possible causes of bias, and the importance of randomizing the treatments.
2. Build your personal interactome. Write down a list of the people that you know on a page. Now organize them by where you tend to meet them, for example at home, school or college, socially, at clubs, and so on. Assign yourself as a node and build an interactome on a sheet of paper using the definitions given in this chapter. Can you identify hubs, degrees, shortest paths? Are there clusters, network motifs, and topological communities? You can assign arbitrary ranking scores to the degrees which connect you to different people. Try using a non-parametric test to identify the links with the greatest statistical significance.

3. The Joint Nature Conservation Committee (JNCC) advises the government about nature conservation. The *National Vegetation Classification Users Handbook* describes sampling methodologies to describe the vegetation which is unique to habitats in the United Kingdom. The handbook is freely available online with a quick web search. Try one of the sampling methods described in the handbook to describe a habitat near to where you live. You might want to compare dandelion distribution in a grass lawn compared to a nearby gravel driveway, moss species in either side of an old east–west running wall, or rowan tree distribution in a heathland habitat above and below a 200m contour line. Try to use the parametric methods described in this chapter to find out if species are restricted to certain habitats.

Further reading

Statistics How To Website.

This website provides explanations, statistical tables, and online calculators for many of the tests discussed in this chapter.

'The t-Distribution.' A Statistics Knowledge Portal. A free online intro to statistics. Jmp Statistical Discovery.

This website provides a clear description of the t distribution and its uses.

Wadhwa, RR and Marappa-Ganeshan, R. *T Test*. Stat Pearls. National Library of Medicine website.

An overview of the assumptions and uses of t-tests in clinical science.

Parametric versus non-parametric: demystifying the terms by Tanya Hoskin. This online article can be found with a quick search.

A description of parametric and non-parametric data sets and the statistical tests that can be used to explore them.

Omics: Exploring space through you. NASA Video channel. YouTube Video.

An inspiring introduction to a series of eight short videos from NASA about the use of omics to understand human biology and the future of medicine.

4 SNAKES, FORTUNES, AND STATISTICAL MODELS

Snakes have always held a fascination for people—although as more people around the world live in cities and towns than the countryside, fewer and fewer of us encounter them in everyday life (Figure 4.1). A look through old folk tales tells a rich story about their significance to human culture, symbolizing movement between mythical worlds and bringing healing, fertility, and foresight. It is said that a lick from a snake's forked tongue will change the path of life and destiny. Folk tales like this celebrate the snake as a symbol of wisdom, change, and fortune. In this chapter, we will use snakes to explore the power of statistics as a sort of scientific crystal ball which helps us to make predictions and find hidden data patterns.

Figure 4.1 From the forked tongue of this UK adder (A) to the gorgeous colours of the Wagler's pit-viper in Singapore (B), snakes are found in most areas of the world, but few of us actually see them. Credit: Anthony Short

A B

Correlation and prediction

Correlation introduces us to the use of statistics in fortune telling or, more precisely, prediction! It provides a way to explain if two variables are linked and whether predictions can be made from their association.

The strength of the association between an independent and a dependent variable is measured using **Pearson's correlation coefficient**, r, which ranges from 0 to 1. Where the variables are unrelated, the data form a random scatter and the correlation coefficient is close to zero. As the association strengthens, the scatter decreases and the points begin to align more closely with a straight line. A perfect linear relationship produces a correlation of +1 or −1, where all points fall exactly on the line. Correlation therefore gives us a measure of the strength of association between variables. A correlation of 0.1 or 0.5 means that the independent variable will predict the behaviour of the dependent variable with 10% or 50% accuracy. A plus or minus sign describes the direction of the relationship with positive associations increasing, and negative associations decreasing, with the independent variable (Figure 4.2).

Correlation coefficients are defined by the closeness of fit of the points to a line that runs between them, and not the slope of the line, differences in scale between the variables or the number of data points on the graph.

Calculating Pearson's correlation coefficient

Let's look at how correlation is used to make useful predictions. As towns expand into desert or jungle habitats, herpetologists—biologists who study reptiles—are often asked to assess the local snake populations to evaluate snakebite risk and support antivenom development (Figure 4.3). At the same time, these surveys play a crucial role in snake conservation, helping protect species whose habitats are shrinking due to human encroachment.

For most people, it's the size and not the species of snake which causes alarm. A person reporting a snake in their house is more likely to say 'There's a snake about a metre long in my kitchen!' than to specify what type of snake it is—they probably, very sensibly, won't get close enough to check!

Does this make sense? Can scientists predict how vemomous a snake might be just by its size? Obviously, this is only going to be valid if the snake is actually a venomous species—so we'll consider the situation if the snake in the kitchen is a rattlesnake. Let's look at how correlation can be used to predict if rattlesnake body length is a useful measure of how much venom is stored in the venom glands. Specially trained herpetologists collect rattlesnake venom by encouraging it to bite through a membrane on a cup (Figure 4.4). The amount of venom released can tell us if snake length

Figure 4.2 Correlation describes the relationship between variables and the reliability of predictions that can be made from the strength of the association.

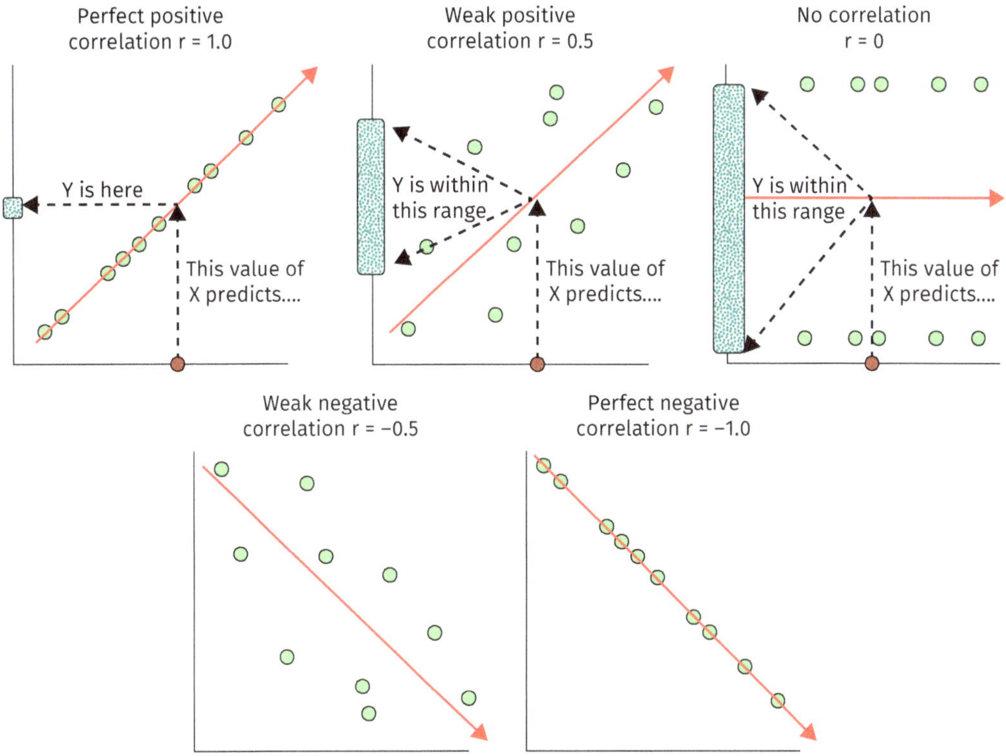

Figure 4.3 Snakes with radiotelemetry tags can be tracked around areas of urban development. In this example from Loma Linda in California, USA, white spots show rattlesnakes found in their natural habitat and red spots show them moving into urban areas (A). As people encroach on snake habitat, more are found in gardens and houses (B). Source: Corbit, AG and Hayes, WK, *Diversity* 14(2), 130, 2022; https://doi.org/10.3390/d14020130 (A)/ Shutterstock 2384886957 (B)

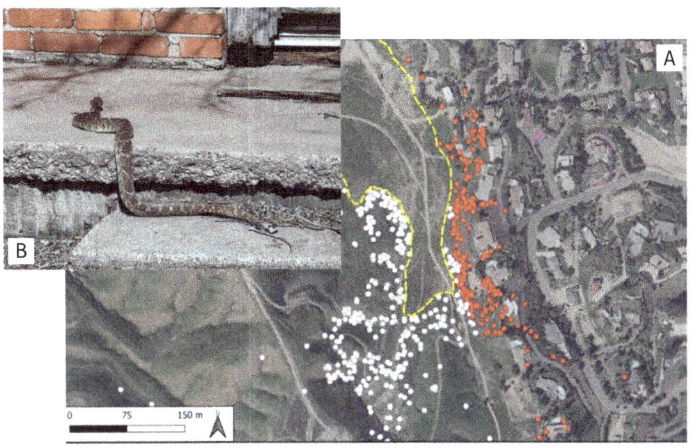

Figure 4.4 Does the length of a snake tell us how venomous it might be? This scatter plot shows the relationship between rattlesnake body length and venom yield.

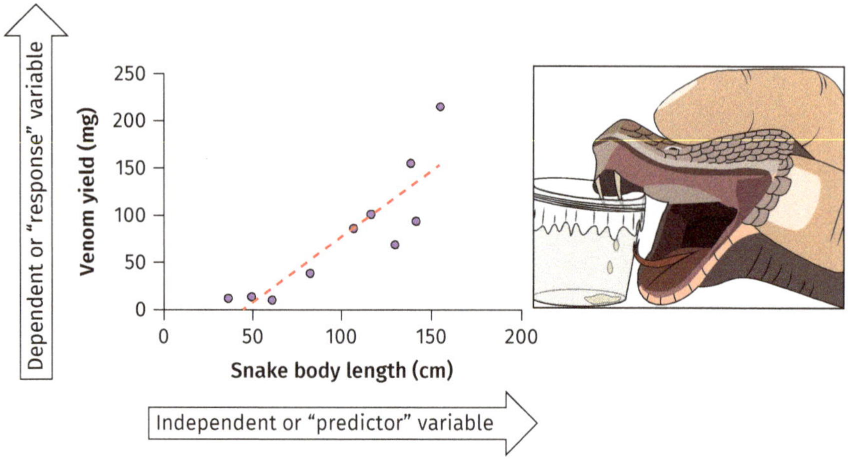

correlates with venomousness. The collected venom isn't wasted—it is used to make antivenom treatments against snakebites. Data from a study of rattlesnakes (*Crotalus* species) using this technique are shown as a scatter graph in Figure 4.4, with the body length as the independent or 'predictor' variable on the x axis and venom yield as the dependent or 'response' variable on the y axis.

$$r = \frac{n\Sigma xy - \Sigma x.\Sigma y}{\sqrt{[n\Sigma x^2 - (\Sigma x)^2].[n\Sigma y^2 - (\Sigma y)^2]}} = \frac{]\text{Signal}}{]\text{Noise}}$$

So, can snake body length be used to predict venom yield? Just like the t-test formula we explored in Chapter 3, Pearson's correlation coefficient, r, measures a ratio of the effect size against randomness in the data. In other words, it measures the ratio of signal to noise, where the differences in the x and y variables on the top of Equation 4.1 above is the 'signal' and variance (standard deviation) on the bottom is the 'noise':

This sum might look complicated, but the different parts of the equation are easily calculated in a table and then inserted into the formula. Figure 4.5 gives the raw data with the calculated terms for signal (red) and noise (blue).

We calculate Pearson's correlation coefficient, r, by placing the terms in Figure 4.5 into the formula, like this:

$$r = \frac{(10 \times 102673) - (1015 \times 794)}{\sqrt{[10 \times 118897 - (1015)^2] \times [10 \times 103328 - (794)^2]}}$$

$$r = \frac{1026730 - 805910}{\sqrt{[1188970 - 1030225] \times [1033280 - 630436]}}$$

Figure 4.5 Calculation of the terms used to determine if snake body length can be used to predict venom yield using Pearson's correlation coefficient.

n	Snake body length (cm) x	Venom yield (mg) y	xy	x^2	y^2
1	36	12	435	1296	144
2	49	14	672	2401	196
3	61	10	624	3721	100
4	82	38	3145	6724	1444
5	107	86	9201	11449	7396
6	130	69	8960	16900	4761
7	116	101	11766	13456	10201
8	141	94	13244	19881	8836
9	138	155	21411	19044	24025
10	155	215	33215	24025	46225
↓	↓	↓	↓	↓	↓
n	Σx	Σy	Σxy	Σx^2	Σy^2
10	1015	794	102673	118897	103328

$$r = \frac{220820}{\sqrt{6.39 \times 10^{10}}}$$

$$r = \frac{220820}{\sqrt{6.39 \times 10^{10}}}$$

$$r = \mathbf{0.87}$$

The correlation coefficient of 0.87 tells us that there is a strong, positive relationship between snake body length and venom yield. Put another way, snake length can be used to predict venom yield with 87% accuracy, but how reliable is this as a predictor of snake venomousness? The next step in correlation is to use a variation of the t-test to find out if the correlation is significant. Here is Equation 4.2 and its workings with the snake venom yield-to-size data:

$$t = r * \sqrt{\frac{n-2}{1-r^2}}$$

$$t = 0.87 * \sqrt{\frac{10-2}{1-0.76}}$$

$$t = 0.87 * \sqrt{\frac{8}{0.24}}$$

$$t = \mathbf{5.0}$$

Remember that statistical and maths formulas often use a dot or asterisk to replace the multiplication symbol, '×'. This is so you don't confuse the multiply command with any value that might be called 'x' or the sample mean, '$\bar{x}$'.

Entering this t value of 5.0 into a t-table at n - 2 = 8 degrees of freedom, and assuming a two-tailed distribution, gives $p < 0.001$. This tells us that we would encounter a Pearson's correlation coefficient of $r = 0.87$ by chance less than one in every 1000 occasions that snake venom yield is plotted against snake length. For anyone with a rattlesnake in their kitchen, snake length really does matter!

Non-parametric correlation using Spearman's rank correlation, r_s

It is sometimes necessary to explore correlations using categorical data, or where the scaling on the x and y axes is too different to allow a Pearson's correlation. As with the non-parametric t-tests in Chapter 3, a ranking procedure can be used to measure correlation based on the distance between data points.

Here is an example. Researchers in the Indian province of Tamil Nadu want to know if annual rainfall patterns can be used to predict the number of people bitten by snakes in villages through the region. A household survey of snakebite incidence was conducted, and this information was linked to monthly rainfall records through the year. The results are shown in Figure 4.6.

Monthly snakebite incidence is categorical because it uses data bins for each month of the year. When plotted against monthly rainfall, it looks as if there is a positive link because data in the wettest months of September, October, and November make the slope through the data points much steeper. Can we really say there is evidence of increased snakebite risk in wetter months?

Correlations using categorical variables can be explored by organizing the data into ranks just as we saw for the non-parametric t-test. This is known

Figure 4.6 Monthly snakebite incidence (bars) and monthly rainfall (red line) in the province of Tamil Nadu, India in 2010. A scatterplot of the same data is shown on the right with monthly rainfall as the independent (x) variable and monthly snakebite incidence (y) as the dependent variable. Data adapted from Vaiyapuri et al., *PLoS One* 8(11): e80090, 2013. doi: 10.1371/journal.pone.0080090. eCollection 2013.

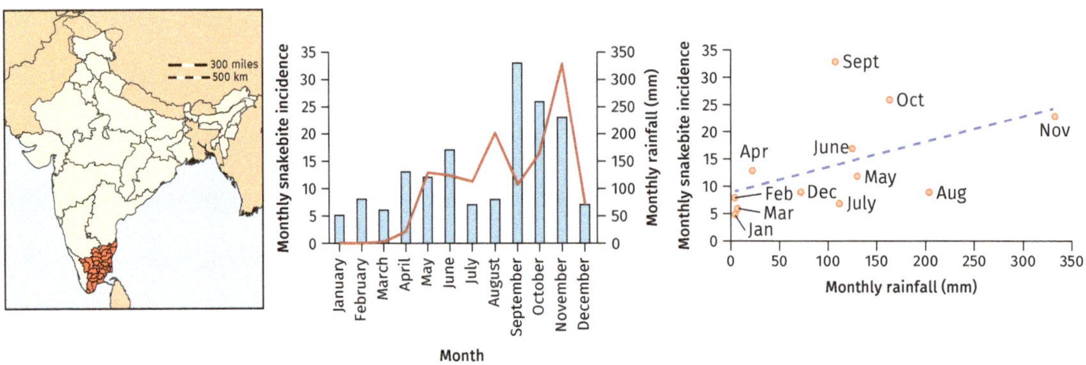

as **Spearman's rank correlation, or r_s**, which uses the formula of Equation 4.3, where d is the distance between the ranked variables:

$$r_s = 1 - \frac{6\Sigma d^2}{n^3 - n}$$

Spearman's rank correlation coefficient is calculated by first ranking each data point—in this case for monthly rainfall and snakebite incidence—from lowest to highest. d is found by subtracting the rank of the dependent variable (y)—the number of people bitten by snakes—from its corresponding independent (x) rank—monthly rainfall. By squaring the distance between values we get rid of negative signs. Table 4.1 shows the calculation of the ranked terms which you then apply to the formula:

$$r_s = 1 - \frac{6\Sigma d^2}{n^3 - n}$$

We can now calculate Spearman's rank correlation coefficient, r_s, by placing the terms in Table 4.1 into the formula:

$$=> 1 - \frac{6 * 126}{12^3 - 12}$$

$$=> 1 - \frac{756}{1716}$$

$$=> 1 - 0.441$$

r_s = **0.56**

Table 4.1 Spearman's rank correlation coefficient.

Month	n	Monthly Rainfall (x, mm)	Rank (x)	Snakebite Incidence (y, people)	Rank (y)	d (x-y)	d^2
Jan	1	5	11	5	12	-1.00	1.00
Feb	2	4	12	8	9	3.00	9.00
Mar	3	7	10	6	11	-1.00	1.00
April	4	22	9	13	5	4.00	16.00
May	5	130	4	12	6	-2.00	4.00
June	6	125	5	17	4	1.00	1.00
July	7	111	6	7	10	-4.00	16.00
Aug	8	203	2	9	8	-6.00	36.00
Sept	9	107	7	33	1	6.00	36.00
Oct	10	163	3	26	2	1.00	1.00
Nov	11	332	1	23	3	-2.00	4.00
Dec	12	72	8	10	7	1.00	1.00

The Spearman's rank correlation value of 0.56 tells us that there is a moderate correlation between rainfall and snakebite incidence—but is this due simply to chance? As with Pearson's correlation coefficient, the r_s value can be used to calculate a t-value as follows:

Entering this t value into a t-table at n – 2 = 10 degrees of freedom and assuming a two-tailed distribution, gives p = 0.06. This tells us that we cannot exclude the possibility that this association has arisen by chance, and so monthly rainfall is a weak predictor of snakebite risk in Tamil Nadu.

Bigger picture 4.1

Correlations and conservation

If you are a tasty, harmless species, you will tend to be easy prey for predators. One way to protect yourself is to evolve skin markings which mimic the warning pattern of a venomous species. This imitation strategy was first documented by a 19th-century Amazon explorer called Henry Walter Bates. He lent his name to the theory of Batesian mimicry, an important line of evidence supporting Darwin's Theory of Natural Selection. Batesian mimicry predicts that predators of edible species which mimic harmful ones must also have knowledge of the harmful species. If this is true, when a harmless mimic moves outside the range of the harmful species, its risk of predation should increase. The harmful species therefore limits the geographical range of the harmless mimic. This has implications for conservation and biodiversity, because harmful species are often persecuted in inhabited areas, but their presence may effectively protect non-harmful species from predation.

Let's look at a classic experiment which tests whether non-venomous snakes gain protection from predator attack by mimicking body colour patterns of venomous snakes. The North American coral snake is a venomous species which lives in tropical latitudes up to 35° north of the Equator and at heights (elevation) of less than 1800 m. Its non-venomous mimic is called the scarlet kingsnake, which lives in the same geographical range.

In this experiment, clay models were made which mimic the ringed colour patterns of coral snakes and kingsnakes (Figure A). By scattering models through habitats that span the latitude and altitude of the coral snake, the researchers can use correlation to predict how the presence or absence of the venomous species influences number of attacks on the kingsnake clay mimics.

The two independent variables are:

- Total attacks on clay replicas.
- Number of venomous snakes in the area.

Figure A Clay models with different colour bands and shades are used to explore how mimicry of venomous snakes by non-venomous snakes protects against predator attack. Top left image is a venomous coral snake; bottom left image is a non-venomous scarlet kingsnake. Photo to the right is of clay models with different colour bands and shades that were used to explore how mimicry of venomous snakes by non-venomous snakes protects against predator attack.

Ai

Venomous

Coral Snake (*Micrurus fulvius*)

Aii

Non-venomous

Scarlet kingsnake (*Lampropeltis triangulum elapsoides*)

Aiii

Clay model replicas of six snake patterns

Figure B uses differences in latitude and height (elevation) to correlate venomous coral snake presence with number of predator attacks. The regression line shows that predator attacks are greatest where coral snakes are not found. Spearman's rank correlation was used to show that coral snakes become increasingly rare at northerly latitudes ($r_s = -0.57$, $p = 0.014$) and at elevation ($r_s = -0.77$, $p = 0.026$) as attacks on ringed clay models increase. The authors conclude that scarlet kingsnake defence by mimicry of coral snakes depends on the abundance of the coral snake itself.

This is important knowledge. Statistical analysis like this tells us that conservation of both species depends on protecting the coral snake from persecution. This same understanding must be applied to conservation projects all over the world—see Case Study 4.1.

Figure B Correlation between latitude (a) and elevation (b) on the proportion of total predator attacks on clay replicas of venomous coral snakes. The blue area indicates number of attacks where coral snakes are present and the green, number of attacks where they are absent. From Pfennig et al., *Nature* 410, p323, 2001.

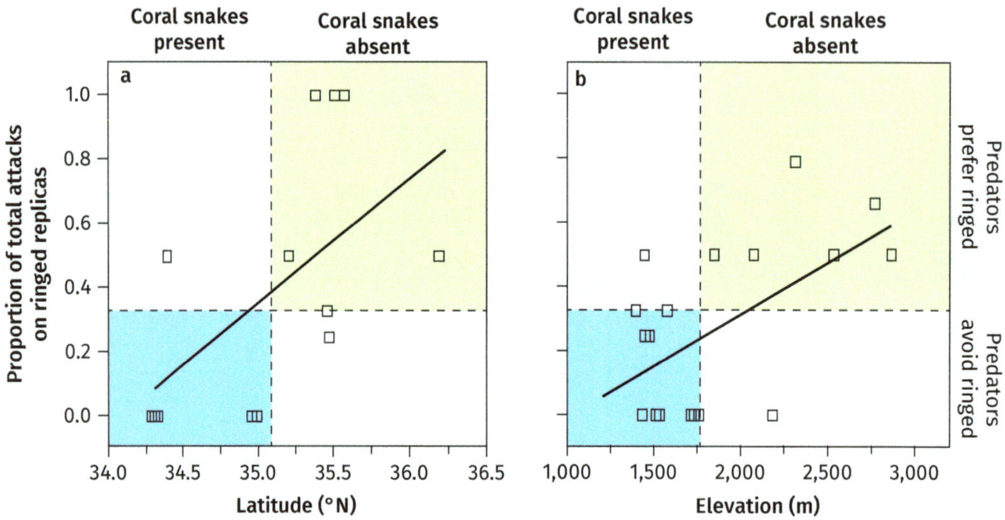

Pause for thought

It is often said that '*correlation is not causation*'.

- Why might it be wrong to say that coral snakes cause scarlet kingsnakes to live where they do? Use the internet to find a distribution for both species in North and South America.
- Are both species always found in the same locations or are there exceptions? Can you think of other factors which might determine where scarlet kingsnakes live?

How linear models predict biological relationships

Correlation helps us to forecast how a dependent variable responds to change in an independent variable, but it tells us little about the variation of the data around the mean of the two data sets.

Linear modelling helps us to explain how data behaves between two variables with different, but constant, rates of change.

Let's go back to Figure 4.5 which showed that snake venom yield and body length are correlated. To explain how closely venom yield varies as body length increases, we first measure how far each point lies from the mean of the response variable, venom yield, to generate a value, **Var(mean)**.

Figure 4.7 Least squares regression calculates R^2, which explains the proportion of the variance in the response variable (venom yield) explained by the predictor variable (snake body length).

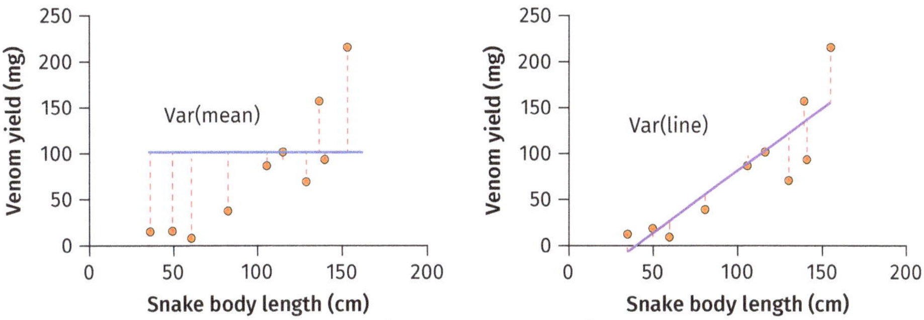

$$R^2 = \frac{\text{Var(mean)} - \text{Var(line)}}{\text{Var(mean)}}$$

Notice that some of the points sit below the mean line and others above, so, to avoid a confusion of negative and positive numbers, all values are squared. Now we fit a line to the data which reports the shortest possible distance from each data point to the line, square it, and obtain the least squared distance, or **Var(line)**. We now have a **least squares regression,** whose slope describes the relationship between the dependent and independent variables (Figure 4.7).

We use the coefficient of determination, R^2, to explain how well the independent variable explains the behaviour of the dependent variable. An R^2 value of 1 shows that 100% of the variance is explained by the regression slope and so we would expect that all points would intersect the regression line perfectly. R^2 values close to 0 indicates that the data is randomly distributed around the regression line. For the right-hand graph in Figure 4.7, the R^2 is 0.76 showing that 76% of the variation in venom yield in snakes is explained by body length.

Confused about the difference between r and R^2? The correlation coefficient, r, gives a measure of the strength of association between two parameters. It allows us to say how well the independent variable predicts the behaviour of the dependent variable. R^2 is a linear modelling term which explains the amount of scatter (variation) around a regression line when the dependent and independent variables are graphed with each other. It's a way to explain how closely the data fits to the line. R^2 is the square of r, the correlation coefficient.

Analysis of variance (ANOVA)

So far, we have discussed how correlation and linear modelling can be used to describe relationships between dependent and independent variables, but most science experiments measure more than just two things. How does linear modelling help us to describe data variation between three or more factors? Analysis of variance, or **ANOVA**, is a type of linear modelling which uses the sum of squares approach to measure the ratio of variation between and within groups. Just as with the t-test, this is a measure of signal to noise, where:

The signal is the variance **between** the groups:

- this is similar to Var(line) which asks '*is there a significant difference in the spread of the data between treatments that might indicate a change or effect?*'

The noise is the variance **within** the groups:

- similar to Var(mean) which asks '*does the spread of data within groups mask the likelihood of any between-group change?*'

$$F = \frac{\text{Variation between treatments (signal)}}{\text{Variation within treatments (noise)}} = \frac{MS_{\text{Treatment}}}{MS_{\text{Error}}}$$

Variation is calculated as the sum of the squared distance to the mean, or **mean square (MS)**, both between treatment groups (we call this $MS_{\text{Treatment}}$) and within groups (we call this MS_{Error}).

In Chapter 3, we saw how the distribution of the t-statistic is used to find out if one variable is statistically different to another. ANOVA uses the ratio of $MS_{\text{Treatment}}$ to MS_{Error} to calculate an **F-statistic** whose distribution can tell us if several experimental groups are jointly different to a control.

The bigger the F-statistic number, the greater the difference between $MS_{\text{Treatment}}$ and MS_{Error} and so the greater the likelihood that differences between the groups did not occur by chance. The F-statistic therefore measures the probability of non-random effects in the data.

ANOVA is often called an **omnibus test** because it tells us if a significant difference exists between the treatments but it doesn't tell us exactly where the differences are found. If the ANOVA produces a statistically significant result, we need to run a follow-up, or **post hoc** test, to find out which of the treatment groups are significantly different to each other. You can think of ANOVA as working in two stages:

- an omnibus test to see if the F-statistic leads us to reject the null hypothesis, followed by
- a post hoc test to see which of the test groups are responsible for this effect.

Why can't we run a Student's t-test on more than two groups?

It's all about the risk of type 1 error (the probability of observing a false positive). Remember that a significance α level of 0.05 means that we would expect 1 in every 20 tests to be found significant simply by chance. As the number of t-test comparisons between treatment groups increases, so the risk of type 1 error increases. We can show this in action using the formula for calculating the risk of type 1 error: $1-(1-\alpha)^n$, where $\alpha = 0.05$ and n is the number of tests.

- Student's t-test always performs only one test to compare two samples.

  $1-(1-0.05)^1 = 0.05$. Here, the risk of type 1 error remains at the critical threshold, or $p = 0.05$ so we're OK.
- Now let's imagine we run three t-tests to compare a control against four sample groups:

  $1-(1-0.05)^3 = 0.1426$. The risk of type 1 error has increased to about 1 chance in 7.
- And five tests for control against six samples:

  $1-(1-0.05)^5 = 0.2262$. Now it has increased to about 1 chance in 4.

ANOVA post hoc tests control for this effect across several treatment groups by keeping the risk of type 1 error at less than or equal to 5%, or 1 chance in 20.

Putting ANOVA into action

It's important to try an ANOVA calculation by hand so that you understand how the process works. Later, we'll do the same calculation using a computer and, although this helps us to be much more rigorous about data assumption checks, significance levels, and measures of effect size, the calculations are hidden behind the scenes. If you first practise by hand, like a car mechanic, you'll understand how things work under the bonnet (Figure 4.8).

The Indian saw-scaled viper (*Echis carinatus*) is one of the 'big four' snake species which cause most snakebite cases each year in India. These are treated using antivenom which contains antibodies that bind and neutralize toxic molecules in the snake venom. The effectiveness of the antivenom depends on the composition of the venom used to make the antibodies, but

Figure 4.8 Understanding how things work can be very helpful!

this varies between different regions of India. If we make the antivenom from snakes collected in one region of India, will it be effective against venom from the same species of snake in other distant regions?

Antivenom effectiveness is tested in the laboratory by finding the dose needed to stop venom from clotting blood in a test tube. The study in Figure 4.9 compared the anti-clotting ability of antivenom made using venom from saw-scaled vipers in Tamil Nadu with its effectiveness against venom from the same species collected in the distant regions of Goa and Rajasthan. The null hypothesis was that the antivenom made from saw-scaled vipers in Tamil Nadu was equally effective against the venom of saw-scaled vipers from Goa and Rajasthan. This is expressed as follows:

$$H_0 : \mu_{\text{Tamil Nadu}} = \mu_{Goa} = \mu_{\text{Rajasthan}}$$

Figure 4.9 Antivenom produced from Indian saw-scaled vipers caught in Tamil Nadu (green) seems to be more effective against snakebites from the same species in the distant regions of Goa and Rajasthan. ANOVA can tell us if this effect is statistically significant. Adapted from Bhatia et al., *Toxins*, 14: 481, 2022 to illustrate the ANOVA process.

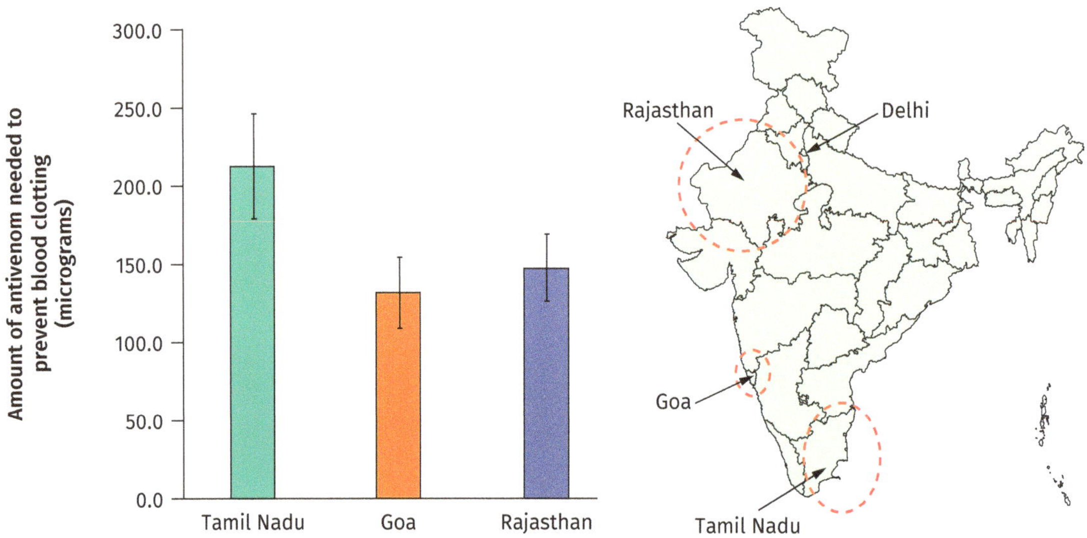

We can see from the graph that the Tamil Nadu antivenom neutralizes poison from saw-scaled vipers from Goa and Rajasthan at a lower concentration, but the whisker bars tell us that the data varies differently within each of the treatment groups.

We can use ANOVA to determine if the differences between each group are random or a predictable effect.

The data is shown in Figure 4.10. 'Between-group' variance, $MS_{Treatment}$, is indicated by the red arrows and 'within-group' variance, MS_{Error}, by the blue arrows. The sample mean of each treatment is shown at the bottom of each column.

Four pieces of information are needed to calculate F-values for the treatment groups:

1. Sum of squares (SS)
2. Degrees of Freedom (DF)
3. Mean of Squares (MS)
4. The mean of all data, or grand mean (GM)

Use the data in Figure 4.10 to work through the ANOVA process in the following six steps:

Step 1: Calculate the between-treatments variation, mean squares for treatments, $MS_{Treatment}$. Begin by subtracting the mean of the Tamil Nadu, Goa, and Rajasthan columns from the grand mean of all data. Square these values to remove negatives and sum together to obtain the sum of squares for the treatments ($SS_{Treatment}$). Now multiply by 6, the sample size of each treatment. Degrees of freedom are called ν_1 and are calculated as the number of treatment groups minus 1. $MS_{Treatment}$ is calculated as $SS_{Treatment}$ divided by degrees of freedom, ν_1 (Figure 4.11 step 1).

Figure 4.10 Use this data table to follow the ANOVA calculation steps. Data have been adapted from Bhatia et al., *Toxins* 14: 481, 2022, to illustrate the ANOVA process.

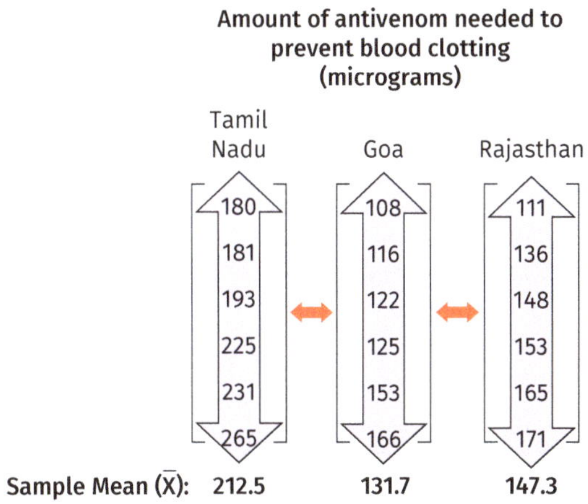

Figure 4.11 Flashcards for each of the first three steps in the ANOVA calculation.

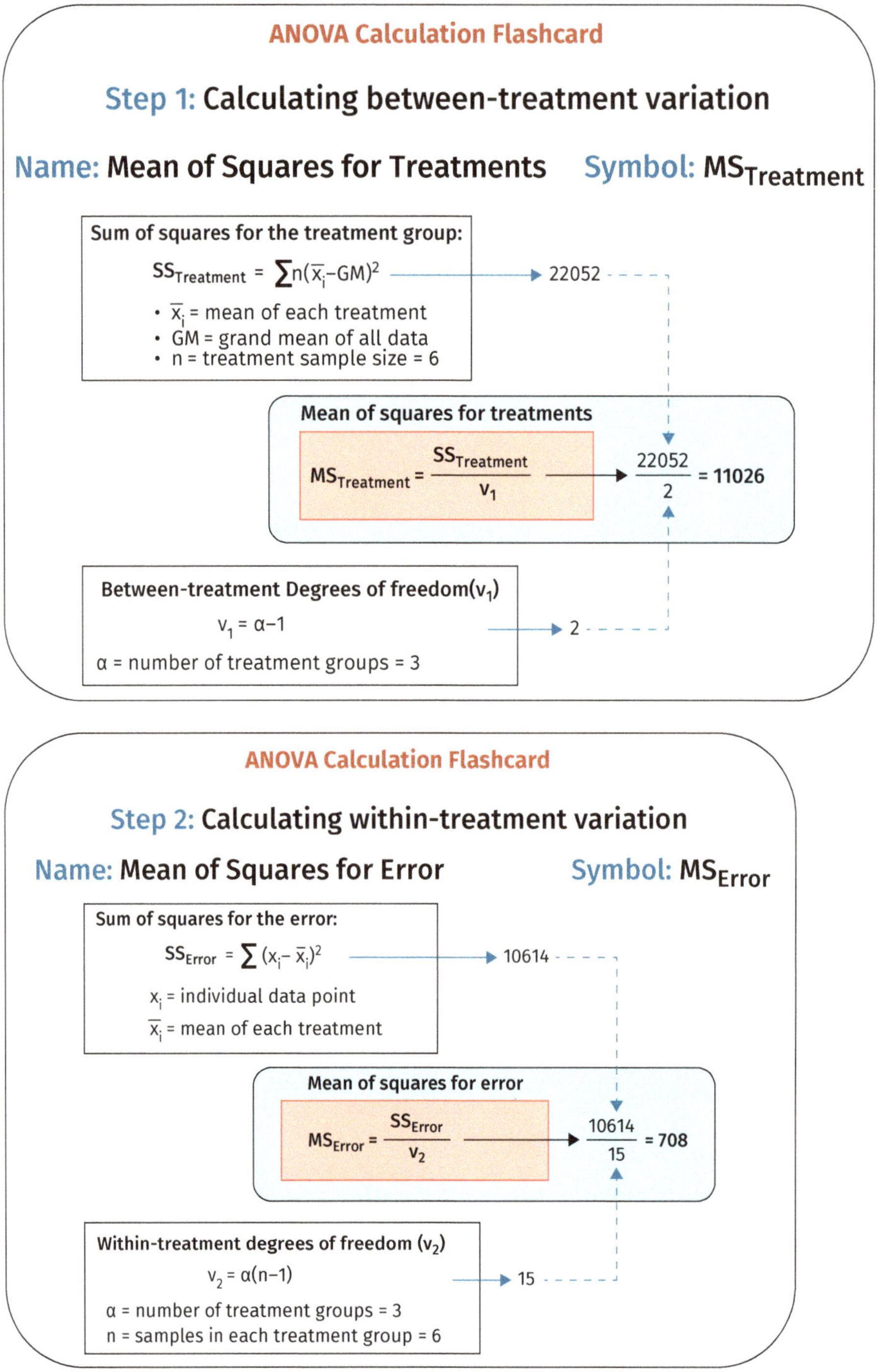

Figure 4.11 *(Continued)*

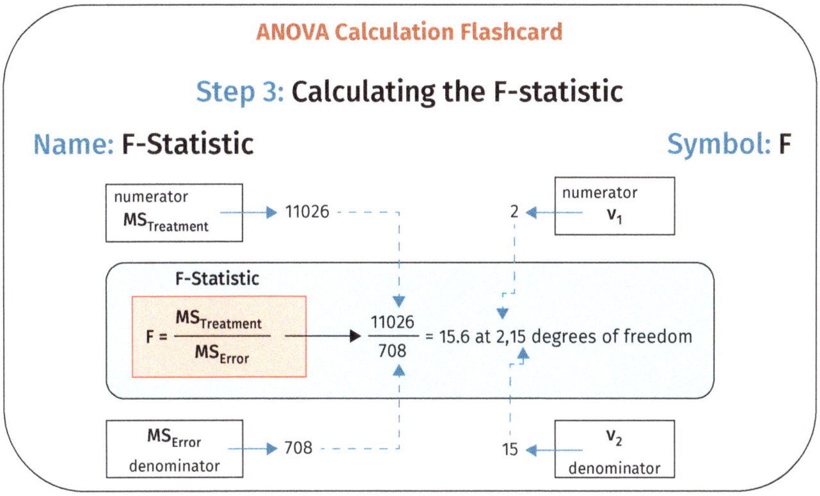

Step 2: Calculate the within-treatment variation, mean square of error, MS_{Error}. Calculate the mean of each of the Tamil Nadu, Goa, and Rajasthan columns. Now subtract each data point from its corresponding mean and square the values. Add the values together to obtain the sum of squares for the error, SS_{Error}. Degrees of freedom for MS_{Error} are called ν_2 and are determined by the number of treatment groups (3), times the number of observations in each group minus one (6–1). MS_{Error} is calculated as SS_{Error} divided by ν_2 (Figure 4.11 step 2). To calculate the total sum of squares, add $SS_{Treatment}$ and SS_{Error} together. Total degrees of freedom are the sum of $\nu_1 + \nu_2$.

Step 3: Calculate the F-value by dividing $MS_{Treatment}$ by MS_{Error}. The degrees of freedom for the F-statistic are determined using ν_1 and ν_2, with the numerator on the top line of the equation given first followed by the denominator on the bottom line of the equation (Figure 4.11 step 3). These will be used to find the F-statistic in the F-distribution.

Step 4: Finding the critical F-value, F_{crit}, for comparison with our calculated F-value. The F-statistic distribution has a different shape to the t-distribution because the mean square calculation of variance only tests probability in a positive direction (see one- and two-tailed t-tests). F_{crit} is the value of the F-statistic at the threshold α probability and is often $p = 0.05$. F values larger than F_{crit} fall in the region below $\alpha = 0.05$ and so require us to reject H_0. Figure 4.12 shows the shape of the F-distribution for 2 and 15 degrees of freedom.

F-statistic tables are calculated for different degrees of freedom and various α levels. Figure 4.13 shows a segment of an F-statistic table for $\alpha = 0.05$. Reading degrees of freedom for $MS_{Treatment}$ across the top ($\nu_1 = 2$) and MS_{Error} ($\nu_2 = 15$) down the left side, we determine an F_{crit} value of 3.6823.

Figure 4.12 Probability distribution of the F-statistic at 2 and 15 degrees of freedom. The critical value of F (F_{crit}) at $\alpha = 0.05$ gives an F-value of 3.6823. F-values larger than this require us to reject the null hypothesis.

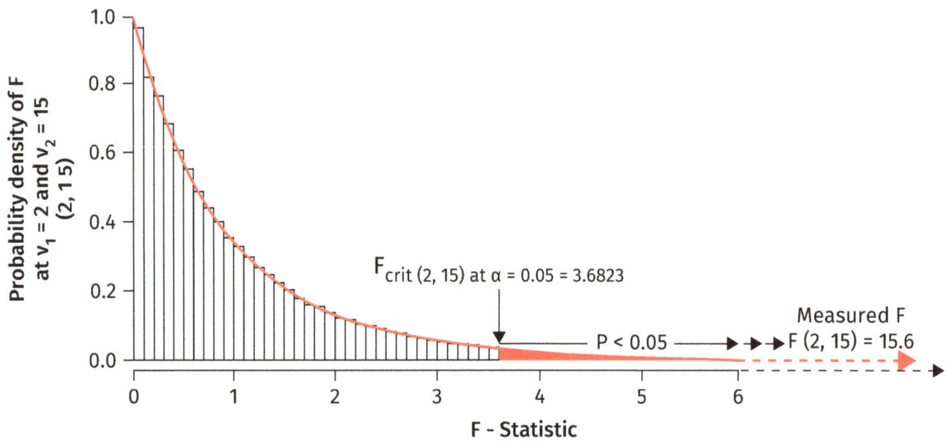

Step 5: Getting the omnibus result. Our calculated value of F = 15.6 is much larger than the critical value, F_{crit} = 3.6823 (at α = 0.05) in the table in Figure 4.13. We can conclude, with more than 95% confidence, that the differences in antivenom potency—when antivenom produced from Tamil Nadu snakes is tested against venom from the same species collected in Tamil Nadu, Goa, and Rajasthan—are unlikely to have occurred by chance. Interesting . . . but which of the three comparisons are significantly different?

Figure 4.13 Table of F-statistic values for 2 and 15 degrees of freedom.

Numerator degrees of freedom, v_1

Denominator degrees of freedom, v_2

	1	2	3
1	161.4476	199.5	215.7073
2	18.5128	19	19.1643
3	10.128	9.5521	9.2766
4	7.7086	6.9443	6.5914
5	6.6079	5.7861	5.4095
⋮	⋮	⋮	⋮
13	4.6672	3.8056	3.4105
14	4.6001	3.7389	3.3439
15	4.5431	3.6823	3.2874
16	4.494	3.6337	3.2389
17	4.4513	3.5915	3.1968
18	4.4139	3.5546	3.1599

Step 6: Post hoc (after-the-fact) significance testing. The Bonferroni α Adjustment Method is a simple post hoc test which works by dividing the critical level of significance used in the ANOVA (α) by the number of groups to be compared:

Bonferroni α = ANOVA α / number of groups in the test.

In our antivenom example ANOVA,

Bonferroni α = 0.05 / 3 = 0.017

A pairwise t-test with Bonferroni correction between the groups is run with a critical α level of 0.017 and gives the following result:

Tamil Nadu:Goa p = 0.0003

Tamil Nadu:Rajasthan, p = 0.00213

Goa:Rajasthan, p = 0.972

We conclude that $p < 0.017$ when clotting efficiency of antivenom prepared from Tamil Nadu snakes is compared to venom from Goa and Rajasthan snakes. There is no statistical difference in clotting efficiency in venoms from Goa and Rajasthan.

Bonferroni's correction increases the risk of type 2 error because it uses an arbitrary approach to raise the critical level of significance. Other post hoc tests such as Fisher's least significant difference (LSD) or Tukey's honestly significant difference (HSD) take account of the type and distribution of the data to minimize the risk of type 1 and 2 errors.

Running ANOVA with post hoc tests using computers

Statistical measurements are often too complex to carry out by hand, so computers are used to do the hard stuff for us! Take care though—it is always important to understand the principles, assumptions, and method of any statistical analysis, whether done by hand or computer. Never accept a p value or statistical end-point unless you know how it was worked out.

Scientists have developed free statistical software packages which can be used by anyone with access to a personal computer. Two common examples are R and JASP. R is a programming language for doing statistics and visualizing data which is often used by people with an advanced interest in statistical programming and bioinformatics. JASP uses the R language but with a user-friendly interface and is what we will use here. Now that you have done the ANOVA by hand, see how JASP helps us to rigorously test the data, starting with assumption checks about normality and variance and ending with a post hoc analysis. Like a car mechanic, now you have an idea about how the engine works! You can download JASP from https://jasp-stats.org/

Use the data in Figure 4.10 to perform the ANOVA test in JASP like this:

Step 1. Import your data into JASP. Use a data management program like Excel to arrange your data into two columns with the independent variable (iv = Region) in column 1 and the dependent variable (dv = Clotting dose) in column 2. Save your file as a comma separated file (filename.CSV). Your data table should look like this:

Import into JASP by clicking the open file tab.

Region	Clotting dose
Tamil Nadu	180
Tamil Nadu	181
Tamil Nadu	193
Tamil Nadu	225
Tamil Nadu	231
Tamil Nadu	265
Goa	108
Goa	116
Goa	122
Goa	125
Goa	153
Goa	166
Rajasthan	111
Rajasthan	136
Rajasthan	184
Rajasthan	153
Rajasthan	165
Rajasthan	171

Step 2. Explore the JASP environment. Once your data is imported, click the 'Descriptives' tab at the top of the screen and the following page will appear. Slide your dependent variable 'Clotting dose' into the 'Variables' box and your independent variable 'Region' into the 'Split' box. You can select the descriptive statistics you wish displayed by scrolling through the options in the lower part of the screen. For now, select boxes to show Valid, Missing, Mean, Standard Deviation, and Shapiro-Wilks. A table of descriptive statistics will appear on the right side of the screen (Figure 4.14).

Step 3. Use the descriptives tab to check assumptions about the data.

- **Assumption 1: Has the data been randomly sampled?** We will assume that the scientists collecting these data were objective and that random sampling was performed. This should be declared in the methodology of a study.

Figure 4.14 The JASP environment lets you set variable and analysis options on the left side of the page and shows results on the right. Different statistical test options run across the top.

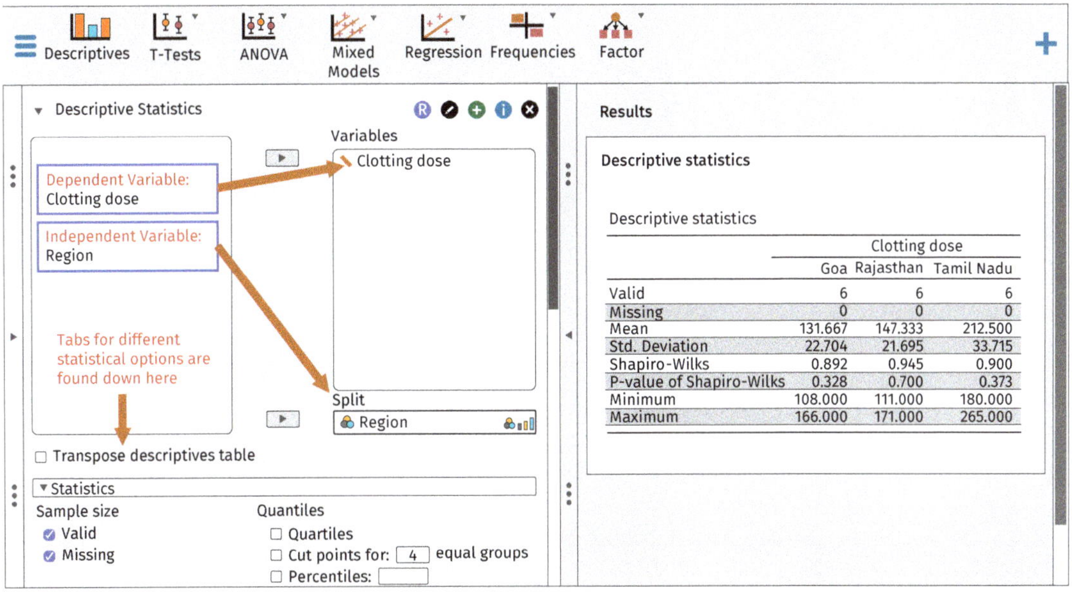

- **Assumption 2: Is the data normally distributed?** The descriptive statistics table shows that the Shapiro-Wilks p-value for each group is greater than 0.05 and so we can conclude all groups are normally distributed. JASP also allows you to check data normality by comparing the quantile distribution of the data in each group with the standard quantile range of a normal distribution. This is called a **quantile–quantile** or **QQ plot**. It's like comparing a cake that you've sliced by hand to one sliced precisely by a computer. If the slices (quantiles) of your data match the sizes of the precisely sliced normal distribution, then all points should align on a graph with a straight-line regression running at 45 degrees. The distance of each data point from the line gives an idea of variance.

To perform a QQ plot in JASP, scroll to the 'Basic Plots' tab at the bottom of the screen and select 'QQ plots'. A QQ plot is generated for each region (Figure 4.15).

The data points for each region sit close to the regression line, and the slopes of each plot are similar, so we can say that the data are normally distributed with similar variance.

- **Assumption 3: Is the variance similar between the groups?** The rule of thumb is that the largest group standard deviation should be no more than double the smallest group standard deviation. The descriptive statistics table shows Tamil Nadu with the largest value (s = 33.7) and Rajasthan with the smallest (s = 21.7) and so this assumption is met.

Figure 4.15 Quantile–quantile (QQ) plots test the normal distribution of data by comparing the quantile range of the data in each group with the theoretical quantile range of a normal distribution (Author (SL)-generated image from JASP).

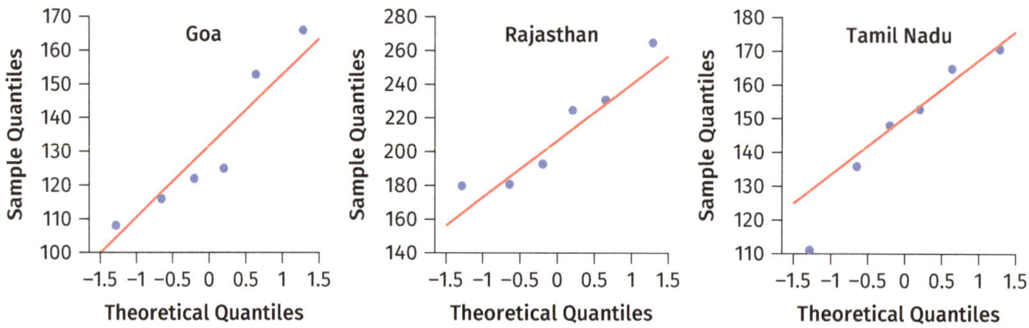

Step 4: With assumptions met, select the ANOVA method. We want to compare clotting dose of antivenom tested against snake venom from three populations: Tamil Nadu, Goa, and Rajasthan. Remember that we must not use a t-test to look for differences between the three populations because the risk of type 1 error increases if the number of comparisons is greater than two. ANOVA controls for this risk and is the right method to see if there are differences between the three populations.

Select the 'ANOVA' tab and set Clotting dose as the dependent variable and Region as the independent variable or 'Fixed Factor'. An ANOVA table is produced (Table 4.2)—do you recognize the sum of square, mean of square, degrees of freedom, and F-value that we calculated by hand?

This is the omnibus ANOVA result. We can see the degrees of freedom and the F-value that we calculated by hand and the probability that the F-value is greater than F_{crit} is $P < 0.001$. There is less than 1 chance in 1000 that we would encounter a false positive in this comparison.

Step 5: Use a post hoc test to assign statistical significance to group comparisons. JASP includes the Bonferroni α Adjustment Method as one of several post hoc test options. Select the 'Post Hoc Tests' tab at the bottom of the page, enter 'Region' into the test box to tell the program you wish to test by the independent variable and click the 'Bonferroni' option. A post hoc table of comparison Bonferroni P values is produced (Table 4.3).

This tells us there is no statistical difference in clotting dose for venoms collected from Goa and Rajasthan ($p = 0.972$) but that the differences between Goa versus Tamil Nadu ($p < 0.001$) and Rajasthan versus Tamil Nadu ($p = 0.002$) are statistically significant.

Table 4.2 ANOVA result table produced in JASP for clotting dose.

Cases	Sum of Squares	df	Mean Square	F	p
Region	2,2052.333	2	11,026.167	15.582	< 0.001
Residuals	10,614.167	15	707.611		

Note: Type III Sum of Squares.

Table 4.3 Bonferroni post hoc results table produced in JASP.

		Mean Difference	SE	T	p_{bonf}
Goa	Rajasthan	-15.677	15.358	-1.020	0.972
	Tamil Nadu	-80.833	15.358	-5.263	< 0.001
Rajasthan	Tamil Nadu	-65.167	15.358	-4.243	0.002

Note: p-value adjusted for comparing a family of 3.

Step 6: How big is the effect? In ANOVA, the effect size indicates how large the differences are between the treatment groups. This is known as eta² (η^2) and is the ratio of the sum of squares for the treatment (this is the Region value in the ANOVA table) divided by the total sum of squares (Region + Residuals from the ANOVA table):

$$\eta^2 = SS_{\text{Region}} / SS_{\text{Total}}$$

$$\eta^2 = \frac{22052}{(22052 + 10614)} = 0.68$$

Interpretation of η^2 varies with the type of ANOVA but, as a rule of thumb,

$\eta^2 = 0.01$ to 0.04 is a small effect
$\eta^2 = 0.05$ to 0.13 is a medium effect
$\eta^2 = 0.14$ or more is a large effect.

Our η^2 value of 0.68 indicates a very large effect that explains most of the variance. We pull together the F-statistic, degrees of freedom, p-value, and η^2 value to report the outcome from ANOVA using ASA style (Chapter 3) as follows:

> 'An analysis of variance showed that the increased efficacy of antivenom produced from Tamil Nadu saw-scaled vipers against venom from the same species found in Goa and Rajasthan was significant, $F(2,15) = 15.6$, $p < 0.001$. The η^2 value was 0.68 indicating a very large effect size. Post hoc analysis by Bonferroni α Adjustment Method indicated that this effect was most significant against venom from Goa ($p < 0.001$), followed by Rajasthan ($p = 0.002$).'

Statistical modelling using several factors

Correlation and ANOVA allow us to compare groups, but they don't reveal other factors that might influence the results. Most of us know that music preferences differ with age. Young people tend to listen to recent music releases and older people often prefer music that spans decades or even centuries. A survey to explore how listening preferences vary with age might compare responses from school-age to retirement-age people about whether they prefer rap or classical orchestral music. The survey produces four treatment conditions, a–d, with results shown in Figure 4.16.

Figure 4.16 A two-by-two comparison of age versus music preference. The differences in the histogram height between treatments a versus b and c versus d suggest an interaction between age and music preference.

	Rap	Orchestral
School-age	a	b
Retirement-age	c	d

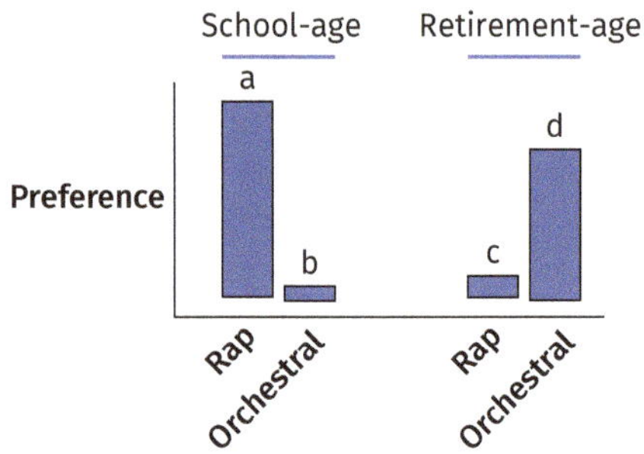

In this example, two factors are at work: an independent variable, age (school age versus retirement age), produces an interaction effect with a dependent variable, music preference (rap versus orchestral). Age, however, might not be the only explanation of music preference. The ability to hear high-pitched sounds (frequencies up to 20,000 hertz) declines as we age and so perception of sound changes through life. Could this affect music preference? We can add low and high hearing range (hr) as a **covariate**. A covariate is a continuous independent variable which can influence the outcome of the main statistical comparison—in this case between music preference and age.

Other covariates can be introduced to produce a **factorial design** experiment (Figure 4.17). These might include physical activity (pa), musical training (mt), personality (py), cultural background (cb), and so on.

We can build these variables into a **maximal statistical model** where each covariate is hypothesized to have some influence on the interaction

Figure 4.17 Measurement of three factors in a factorial design experiment. Each factor has two or more conditions known as levels. Factorial design experiments allow study of the effect of each factor on the dependent variable and allows exploration of interactions between all factors.

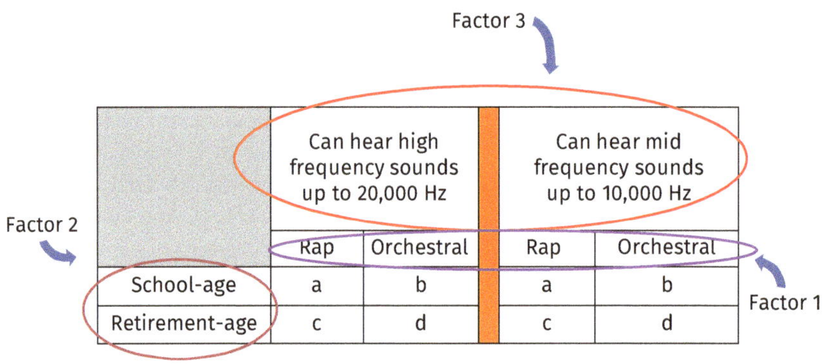

between the measured response and the treatment group. The maximal model for all possible interactions is:

- The measured response (dependent variable, dv): music preference
- The treatment group (independent variable, iv): age
- The covariates (cv): hr, pa, mt, py, cb.

The summary formula for the maximal model is written as:

music preference ~ age × hr × pa × mt × py × cb . . .

Each covariate adds a new level of complexity to the model. In this example, the measure of music preference against age must be tested against every possible combination of the 5 covariates giving 120 interactions!

Analysis of covariance (ANCOVA) allows researchers to understand how covariates influence the relationship between the dependent and independent variable. It uses the ANOVA method to assign F-scores and p-values to each combination of factors in the maximal model. A model simplification process is used to find the fewest possible interactions which explain the interaction between the dependent and independent variables. This is known as the **Minimum Adequate Model (MAM)**.

ANCOVA and model simplification using JASP

Let's return to our snakes to look at how scientists use ANCOVA to simplify statistical models.

Each year about 8000 people are bitten by venomous snakes in the USA and, although rarely fatal, hospital treatment is usually needed. It's helpful for doctors to understand the factors that determine how long a patient will be in hospital. We can build a maximal model which compares time spent in hospital with factors that might affect recovery time, like:

- size of snake (size)
- weight of the patient (weight)
- number of antivenom vials used (vials)
- time elapsed from bite to hospital admission (time).

Figure 4.18 Just like an archaeological dig, ANCOVA uses statistics to brush away noise which covers up underlying data patterns.

Here, length of hospital treatment is the dependent variable and snake size is the independent variable because it gives an important indication of possible bite severity. Patient weight, the number of antivenom vials used during treatment, and time between being bitten and treatment might also affect how long a patient is in hospital and so these are covariate factors.

The graphs in Figure 4.19 show how length of hospital treatment behaves when plotted separately against each covariate for patients bitten by small or large snakes. From a quick look at the scatter of points on each plot, it seems that snake size may well affect length of hospital treatment but so might patient weight, number of vials used to treat the bite, and time to treatment. It's hard to know which of these very different factors are most

Figure 4.19 Scatter graphs showing the relationship between snake size and duration of hospital treatment. Each graph shows how the data behaves if the covariates of patient body weight or vials of antivenom are plotted as independent variables. Source: Data adapted from Corbit, AG, *The Dynamics of Human and Rattlesnake Conflict in Southern California* (2015). Loma Linda University Electronic Theses, Dissertations & Projects, 347.

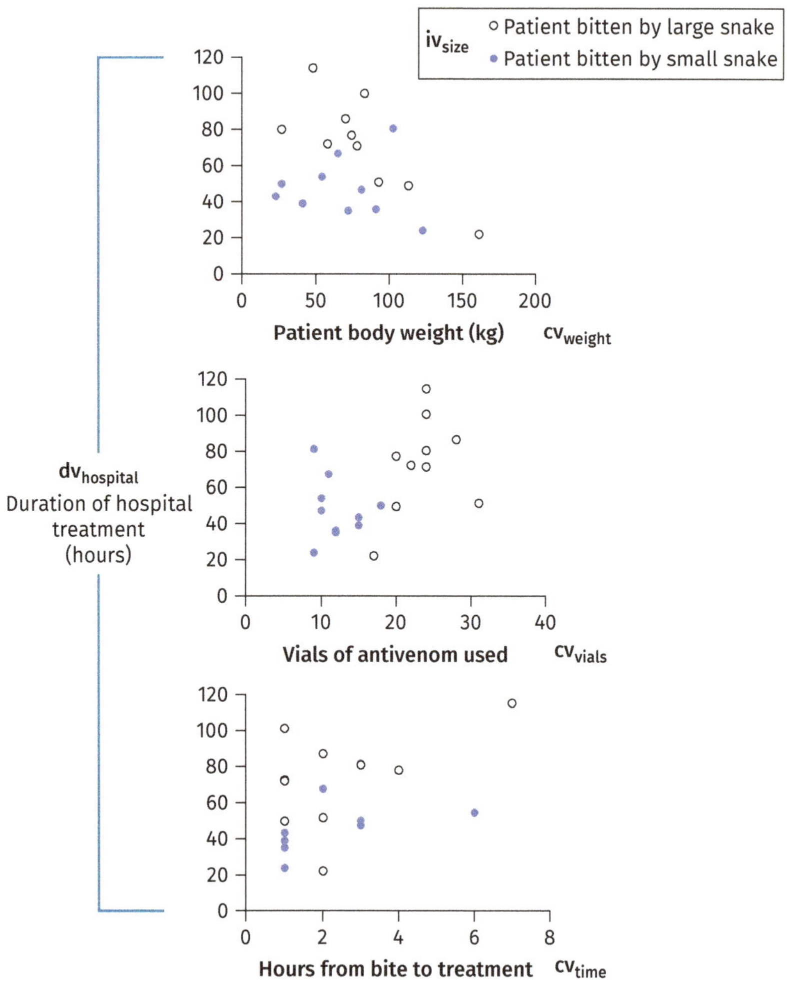

Table 4.4 Duration of hospital treatment for patients bitten by small or large snakes.

iv_{size}	$dv_{hospital}$	cv_{weight}	cv_{vials}	cv_{time}
Size of snake	**Hours in hospital treatment**	**Kilograms of patient weight**	**Number of antivenom vials used**	**Hours between bite and hospital treatment**
small	50	27	18	3
small	43	23	15	1
small	39	41	15	1
small	54	54	10	6
small	67	65	11	2
small	35	72	12	1
small	47	81	10	3
small	36	91	12	1
small	81	103	9	3
small	24	123	9	1
large	80	27	24	3
large	114	48	24	7
large	72	58	22	1
large	86	70	28	2
large	77	74	20	4
large	71	48	24	1
large	100	83	24	1
large	51	93	31	2
large	49	113	20	1
large	22	161	17	2

important for helping doctors predict how long a patient needs to stay in hospital to recover from a snakebite (see Table 4.4 for data for this example). Can ANCOVA and model simplification find patterns in the data that aren't immediately visible on the graphs?

Here is the data from Figure 4.21 which we will use to show the ANCOVA process using JASP.

Step 1. The process starts just like ANOVA. Import your data saved as a comma separated file (CSV). Except for the independent variable, all data is continuous and so select the column heading to make sure there is a 'ruler' symbol next to each column title.

Step 2. Click on the ANOVA tab and select ANCOVA.

Step 3. Slide each variable into its correct box on the page and make sure you click the option below to show effect size (η^2). We have now told JASP about the parameters in our maximal model:

$$dv_{hospital} \sim iv_{size} \times cv_{weight} \times cv_{vials} \times cv_{time}$$

Step 4. JASP produces a table (Table 4.5) which compares our dependent variable, duration of hospital stay, with the independent variable, snake size, and the three covariate conditions of weight, vials, and time. You will recognize all of the terms in this table from the ANOVA earlier in the chapter.

At first glance, our maximal model seems to show no effect among any of the conditions. The F-values are low, p-values are well above the $p = 0.05$ threshold of significance and the effect sizes are very small. Remember that the maximal model includes our dependent and independent variable relationship plus all of the variance produced by the three covariates. Just like the archaeological dig (Figure 4.18), we need to systematically sweep away this noise to see if there are underlying patterns.

Step 5. The table shows that the covariate with the smallest effect size is CV-vials. Let's remove this to see the effect on the other relationships. In JASP, simply drag cv-vials over to the data source box on the left. The ANCOVA analysis automatically refreshes to produce a new table for the refined model (Table 4.6).

$$dv_{hospital} \sim iv_{size} \times cv_{weight} \times cv_{time}$$

By removing the effect of cv-vials we now find that snake size is strongly associated with length of hospital stay. The covariate of cv-time remains statistically insignificant as does patient weight. Can we refine the model further by removing cv-time?

Step 6. Remove cv-time to test the newly refined model:

$$dv_{hospital} \sim iv_{size} \times cv_{weight}$$

The ANCOVA table produced looks like Table 4.7.

Table 4.5 Maximal ANCOVA results table produced in JASP for duration of hospital stay.

Cases	Sum of Squares	df	Mean Square	F	p	η^2
iv – size	755.763	1	755.763	2.168	0.162	0.095
cv – weight	1,016.869	1	1,016.869	2.918	0.108	0.127
cv – time	972.350	1	972.350	2.790	0.116	0.122
cv – vials	19.341	1	19.341	0.055	0.817	0.002
Residuals	5,227.954	15	348.530			

Note: Type III Sum of Squares

Table 4.6 Effect of removing cv-vials on ANCOVA results produced in JASP for duration of hospital stay.

Cases	Sum of Squares	df	Mean Square	F	p	η^2
iv – size	3,462.074	1	3,462.074	10.557	0.005	0.306
cv – weight	1,415.258	1	1,415.258	4.315	0.054	0.125
cv – time	1,174.752	1	1,174.752	3.582	0.077	0.104
Residuals	5,247.296	16	327.956			

Note: Type III Sum of Squares

Table 4.7 Miminum Adequate Model (MAM) ANCOVA results produced in JASP for duration of hospital stay.

Cases	Sum of Squares	df	Mean Square	F	p	η^2
iv – size	3,979.234	1	3,979.234	10.534	0.005	0.312
cv - weight	2,365.952	1	2,365.952	6.263	0.023	0.185
Residuals	5,247.296	16	377.768			

Note: Type III Sum of Squares

This table shows two important things. Firstly, removing cv-time has had very little effect on the independent variable, snake size, and so we seem to have refined the effect of the covariates as far as we can go. Secondly, however, the F-value for the last remaining covariate, patient weight, has increased to produce a significant effect with an increased effect size. The relationship, $dv_{hospital} \sim iv_{size} \times cv_{weight}$ is our **Minimum Adequate Model, MAM**, and it tells us that patient weight significantly explains part of the relationship between length of hospital stay and snake size.

Not so fast! What about our assumptions? Is the data normally distributed with similar variance in our MAM-ANCOVA model? Open the 'Assumptions' tab in JASP and generate a QQ plot. If our data are normally distributed with equal variance, we should find that our data points are close to a perfect regression between the axes.

It looks like that's the case from Figure 4.20. We are now free to report the output from the MAM-ANCOVA table APA style using the degrees of freedom, F-value, p-value, and η^2 value as follows:

> 'An analysis of covariance showed that the independent categorical variable of snake size significantly predicted length of hospital stay, $F(1,17) = 10.534$, $p < 0.01$ with a very large effect size of $\eta^2 = 0.312$. The covariate, patient weight (kg), also significantly predicted the subjects' length of hospital stay $F(1,17) = 6.263$, $p < 0.05$ but with moderately large effect size, $\eta^2 = 0.185$.'

Figure 4.20 An interquartile or QQ plot shows if the data is normally distributed with equal variance.

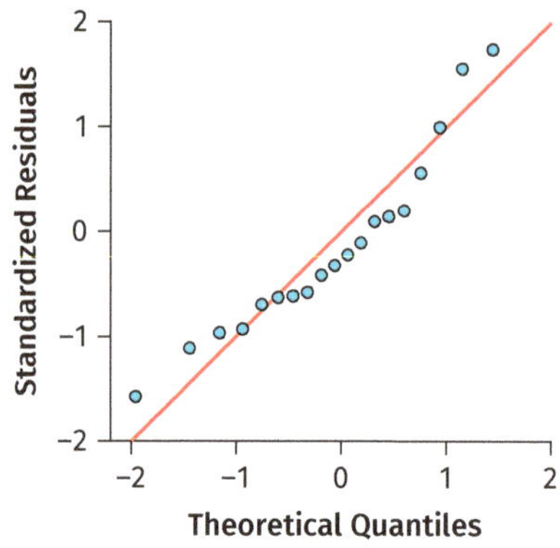

Step 7. Lastly, we want to tell people about this important discovery, but showing numbers and tables might not win you many friends! It's much better to help people see what the relationship looks like by drawing a plot. You can do this in JASP by opening the 'Descriptive plots' tab and placing cv-weight in the 'x axis' box and iv-size in the 'separate lines' box. You should get a graph like Figure 4.21.

It's easy to see that a snake bite from a large snake will result in a much longer stay in hospital for a light person, like a child, compared to a heavy person, like an adult. Length of hospital stay for bites from small snakes seems much less dependent on body weight. If a snakebite victim comes into an emergency treatment centre, a doctor can trust the statistics in this graph to quickly decide the right course of treatment and predict recovery time. Statistical tests like this are the foundation of evidence-based medicine which we met in Chapter 1. In its Guidelines for the Management of Snakebites, the World Health Organization (WHO) say that either a '*snake identified as a very dangerous one or a large specimen*' are the most important first signs that a person has received a serious snake bite. Statisticians at the WHO did our calculation before us!

Figure 4.21 This graph shows the ANCOVA result as a linear model of duration of hospital stay in hours (dv–hospital) and patient weight in kilograms (cv–weight) for patients who have been bitten by large or small snakes.

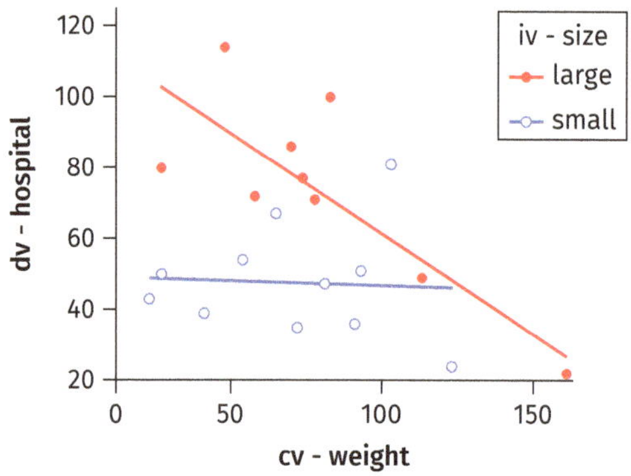

Statistical approaches to complex problems

Life processes are complicated and normally work through a web of connections so we need to build statistical models which can handle lots of interacting things all at once. ANOVA and ANCOVA provide a foundation on which we can add a second independent variable to grow from a one-way to a two-way analysis. If we add a second dependent variable our analysis now tests multiple variables and is called multivariate. Multivariate-ANOVA and ANCOVA (MANOVA and MANCOVA) can be one-way or two-way and each compares two dependent variables against one or two categorical factors.

This sounds complicated—so let's look at an example. Imagine we want to know if lung volume (LV) varies between males and females.

- Biological sex is our first independent categorical variable which we can call 'Factor 1 - sex'.
- We know that lung function declines with age and so our study group is subdivided into two additional groups: young and old. We have now created a second categorical independent factor which we call 'Factor 2 - age group'.
- Now we decide that height could influence LV independently from sex or age group because tall people have larger chest sizes. This is included as a covariate.
- Finally, we are interested to know if heart rate may be associated more closely with sex, age group, and height than lung volume.

Table 4.8 breaks down these comparisons with complexity increasing from top to bottom. You can use a similar table to work out the type of statistical model you should use to test your own experiment

Table 4.8 Characteristics of multivariate statistical models based on ANOVA.

Statistical Test	dv	Categorical variable (iv)	cv	dv	Categorical variable (iv)	cv
1-way ANOVA	1	1		LV	Factor 1 – sex	
2-way ANOVA	1	2		LV	Factor 1 – sex Factor 2 – age group	
1-way ANCOVA	1	1	Yes	LV	Factor 1 – sex	height
2-way ANCOVA	1	2	Yes	LV	Factor 1 – sex Factor 2 – age group	height
1-way MANOVA	2	1		LV	Factor 1 – sex	
2-way MANOVA	2	2		LV HR	Factor 1 – sex Factor 2 – age group	
1-way MANCOVA	2	1	Yes	LV	Factor 1 – sex	height
2-way MANCOVA	2	2	Yes	LV HR	Factor 1 – sex Factor 2 – age group	height

Things are getting complicated!

Permutational multivariate ANOVA (**Permanova** for short) is a type of non-parametric multivariate analysis used to find out if clustered data points on a graph can be classed as a group with shared properties or are randomly scattered. It assigns significance to different arrangements, or permutations, of the data by measuring the distance, difference, and similarity of various possible groupings (Figure 4.22). This makes it useful for analysing qualitative or non-normal data and is often used in ecological, wildlife conservation, microbiological, or genetic studies. Permanova is used to analyse data cluster patterns. It assigns statistical significance to differences found between groups and so is a useful tool for multivariate hypothesis testing.

Figure 4.22 Permanova identifies groups by measuring distance to a central point within a data cluster (dotted lines). Significance is assigned by determining the distance of groups from one another (solid lines).

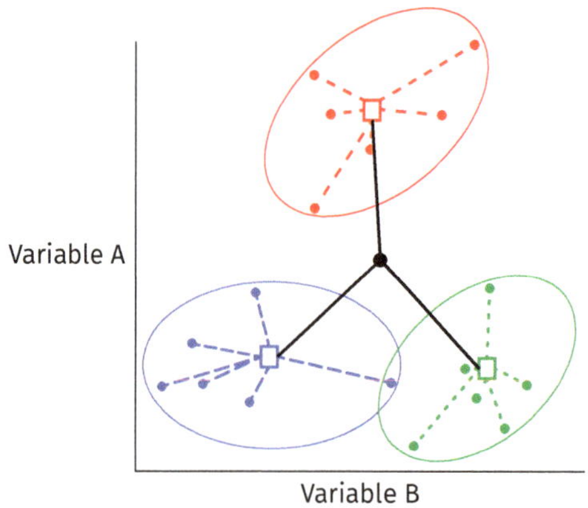

Case study 4.1

How do scientists explain complicated statistical findings to the public?

Wildlife conservation often depends on the goodwill of local communities, but preserving biodiversity means looking after popular and unpopular species (see Bigger Picture 4.1). People in rural communities often have deep feelings about the wildlife in their area, some of which may be shaped by traditions, myths, folklore, and superstitions. Ethnobiologists study these beliefs to build bridges between scientists, government, and communities that are vital for conservation work.

Ethnobiologists in Colombia used Permanova to search for patterns of belief about conservation among people living in rural communities of the Andean mountains. They used a questionnaire to survey attitudes to 17 mammal, bird, reptile, and amphibian species local to the region, which asked people to respond to four statements of their relationship with the animals:

1. I like this species
2. I believe this animal is harmful to my normal activities
3. I think this animal might transmit disease
4. This animal has medicinal benefits.

Responses were scored on a 5-point Likert scale where 1 was totally disagree and 5 was totally agree. The researchers then gathered information about stories, myths, or traditions linked with each species. Finally, people were asked if they would support local government in protecting or eradicating the species.

Figure A Permanova can be used to develop sophisticated public communication methods such as these polygon maps which summarize attitudes to local animal conservation priorities. These images are from Moreno-Rubiano et al., *Ethnobiology and Conservation*, 12:09, 2023, doi:10.15451/ec2023-06-12.09-1-20

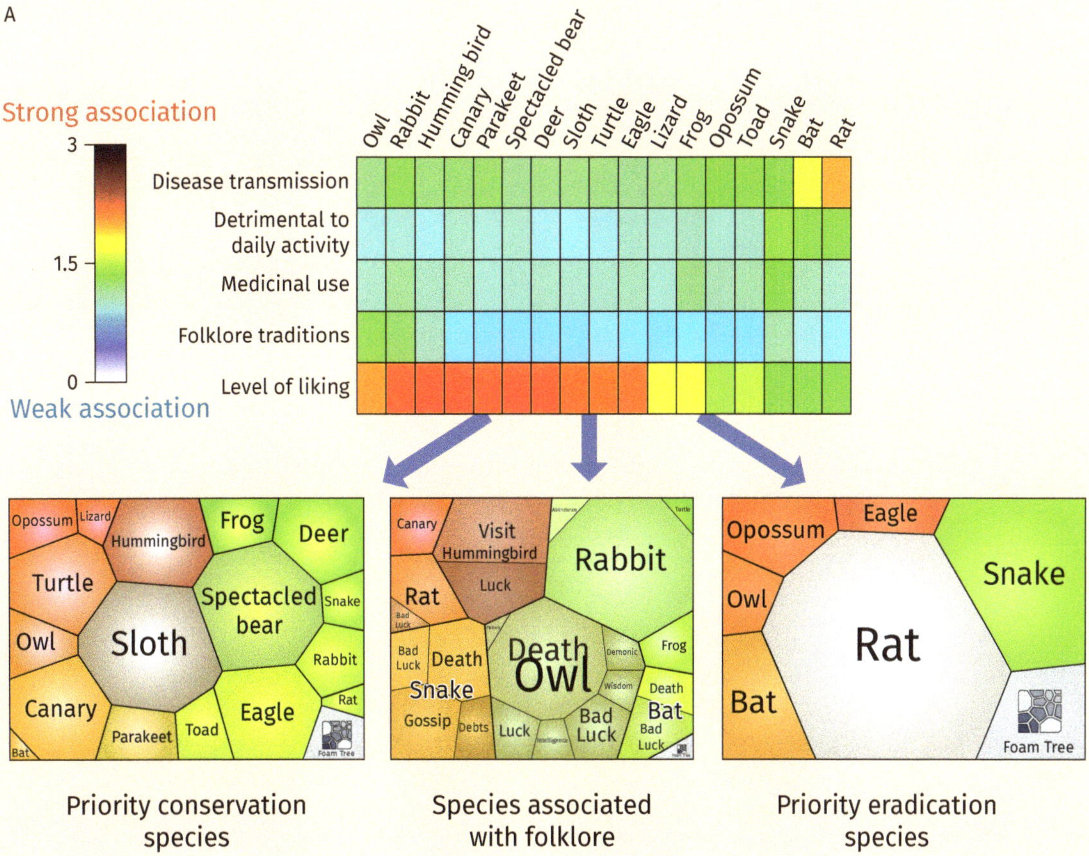

Permanova was selected as the best method for finding clusters in the data that might vary by animal grouping (mammal, bird, reptile, amphibian), geographic group (urban or rural), sex of the person completing the questionnaire (male or female), and their age group (five age categories from seven to 70 years).

The Permanova results were used to create a heat map which shows how each factor was associated with different animals (Figure A). Red colour shows high association, green is neutral, and blue is weak. The results were used to create polygon maps that summarize how perceptions vary for different types of animal.

Species like sloths, spectacled bears, canaries, turtles, eagles, and hummingbirds (Figure B) were most liked and associated with high conservation priority. These might serve as flagship species for conservation efforts. Least-liked species such as rats (Figure C), snakes, bats, and opossums were associated with negative traditions and present a greater challenge for conservation. Here, the approach is to show how preservation of unpopular species is essential for healthy ecosystems. An awareness of the place of these animals in local traditions and folklore gives conservationists a way to teach people about their work in a way that is sensitive to the culture and traditions of local people.

Figures B and C Using statistics to help identify animals people like, such as this hummingbird (Figure B), and ones they are less keen on—like the rat (Figure C)—helps build effective conservation partnerships.

B

C

Pause for thought

Rural communities often provide vital support for conservation work because they understand the delicate balance between human habitation and the natural world. Think of a conservation issue near you, such as repurposing of old industrial land (so-called brown field sites), protection of green belts around urban areas or, perhaps, recovery of marsh land to increase biodiversity.

- What sort of data would you need to present a case to the public supporting your view of how the land should be used?
- How might you study the beneficial effects of unpopular species like wasps, spiders, leeches, or weed plants as natural pest controllers and nitrogen processors? Can you see how statistical techniques might help you to build a scientific foundation for your case?

A look to the future: statistical modelling for artificial intelligence

The goal of artificial intelligence is to create computer systems which think much like a human brain. Statistical modelling of the type we saw in ANCOVA, MANCOVA, and Permanova creates a computer training framework known as machine learning which we encountered at the end of Chapter 2. Here, large amounts of data are provided to the computer, and human intervention is needed to direct the computer algorithm to an optimal solution to a problem.

The future goal is to move to deep learning where computers use electrical neural networks to identify problems, find information, and create solutions without human intervention. In biology, statistical modelling in partnership with AI is used to develop strategies for precision medicine, mine bioinformatic databases, create new drugs for orphan diseases, and develop intelligent automated solutions for industrial biotechnology and agriculture (see Figure 4.23).

Figure 4.23 Statistical modelling helps to develop artificial intelligence solutions which will advance thinking in every area of the biosciences in the future.

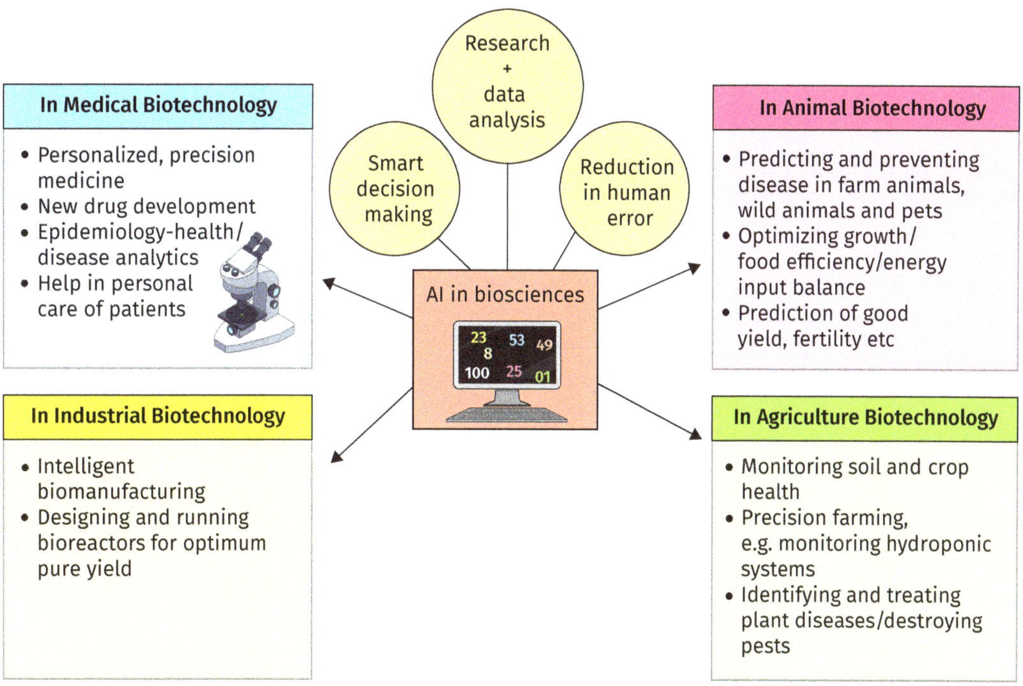

Chapter summary

- Correlation helps us predict associations between two things but be careful, it does not imply that they cause one another.
- Linear models use least squares regression to explain how data behaves between two variables with constant rates of change.
- Analysis of variance (ANOVA) is a linear modelling technique which helps us measure differences in variation between three or more comparison groups.
- ANOVA works as an omnibus test which uses the F-distribution to find if a statistical difference is present somewhere between the groups. A post hoc test is needed to find which groups are statistically different to one another.
- As statistical tests become more complex it becomes helpful to use a computer to perform the calculations. As a careful and objective scientist, it is always important to understand the principles, assumptions, and method of any statistical analysis. Never just let the computer do the work!
- Analysis of covariance (ANCOVA) is a type of statistical model which takes account of several factors, called covariates, which might affect the relationship between the dependent and independent variables.
- ANCOVA provides a foundation for complex statistical models which look for data patterns among multiple variables, such as MANOVA, MANCOVA, and Permanova.
- These complex models are the basis for artificial intelligence and machine learning technologies.

Discussion questions

1. Occam's Razor (often called the principle of parsimony) says that the simplest explanation of a problem is usually the best. How might covariates add complexity to a statistical model? Do you think it's fair to always assume that the simplest explanation is the best?
2. 'Numbers never lie' and 'It's easy to make the numbers show whatever you want' are conflicting statements that are sometimes mentioned when statistics are used in debate. Do you think misinterpretation of causation and correlation might fuel these arguments? You might like to look at some examples of unexpected correlations in the correlation.org website to start your own debate.
3. Discuss the ways in which statistics are used to guide us through daily life. Think about the last time you followed advice about things like health, food preferences, or safety that may have had their origins in statistics. Compare how simple or complicated you think the statistical analysis might have been with the way it was communicated to you.
4. In this chapter we looked at ways that correlation, ANOVA, and statistical modelling can be used to predict outcomes and assign

significance to covariates. Here are two exercises which will help you to explore the power of statistical modelling close to home.

a. Buddleia (the butterfly bush), rosebay willow herb, cow parsley, brambles, and clover are a few examples of common nectar-rich or berry-producing flowering plants which often grow in untended margins of parkland but how important are they for biodiversity? Conduct a conservation project in miniature by recording the number of flowering plant species you can find in a defined area (for example, 25 paces long by 1 pace deep if the margin allows). Record the number of insect, bird, and mammal species associated with the area. A sweep net made from fine mesh muslin can help to collect insects and the presence of birds and mammals can be observed directly or indirectly by looking for signs like droppings or burrows. Carry out the same species survey on a neighbouring patch of tended parkland such as a lawn or (with permission and care) a tended flower bed with about the same area as your sampled margin. You should repeat this exercise in different areas of the park to increase the statistical power of your sample. Plot a graph with the number of nectar-rich/berry-producing plants as the independent variable and the number of insect, bird, and mammal species as the dependent variable. Can you describe the relationship between the two using correlation? How does the number of times you sample affect the correlation coefficient, r? How do other factors like weather conditions, sunlight versus shade, or time of year affect this relationship? How might you use this information in a poster to improve public awareness of the importance of biodiversity?
b. We can look at drought-resistant crop variants using ANOVA. For example, cress seeds normally need damp conditions to grow but varying the amount of water can reveal genetic variants which might be drought resistant. For this to work, it is important to buy a packet of fresh seeds—don't use one that has been at the back of a drawer in the school lab or at home for several years! Place 50 cress seeds in a 10×5 grid in the middle of a tray of compost. Repeat with two more trays keeping the arrangement of the seeds the same. Cover them with a thin layer of grit or sand and water all 3 trays at the start of the experiment with 50 ml water sprinkled over the surface of the seeded area. Note this as 'day 0' in a lab book. Now label tray one as '1. Water every 2 days', tray 2 as '2. Water every 4 days' and tray 3 as '3. No further water'. Record the number of seeds which have germinated with each watering regime each day up to 14 days. Draw a scatter graph for each tray and use the ANOVA protocol in JASP to determine if your data is normally distributed and if the F-statistic is statistically different between tray 1 compared to trays 2 and 3. Use the post hoc testing protocol to find which of the trays are different. What further experiment might you do to find out if the seedlings which germinate in tray 3 are truly drought resistant?

Further reading

BBC Bitesize series: *How are scientific explanations developed?*—OCR 21st Century. Online site.

An introduction to the terms, assumptions, and uses of correlation in biology.

ANOVA (Analysis of Covariance). Biology for Life. Website.

Examples of Correlation and ANOVA methods plus other statistical tests, aligned with the International Baccalaureate syllabus.

Field A. Discovering Statistics: Analysis of Covariance (ANCOVA). Discovering Statistics website. PDF document.

A more detailed look at the history, assumptions, and method for performing ANCOVA and related analyses.

Spiegelhalter, D. *The Art of Statistics: Learning from Data*. Pelican Books: 2019.

A discussion of the ways in which complex statistics are shared and communicated clearly with the public.

Mullin SJ and Seigel RA. *Snakes: Ecology and Conservation*. Cornell University Press. 2009.

Describes the challenges and implications of snake conservation with examples of the role of statistics in snake conservation.

5 BIAS AND THE SCIENCE OF EPIDEMIOLOGY

Everyone is different—even identical twins! There is a good chance that, out of every 10 people who read this chapter, eight won't like Brussels sprouts, three will describe blue as their favourite colour (four if you're male), seven will prefer dogs to cats, and most will have strong but widely different opinions about the latest mobile phones on the market. When it comes to our view of things, we are all inherently, unavoidably, beautifully biased!

Bias makes us who we are. It gives shape to our opinions, preferences, choices, sense of cultural identity, and beliefs. Bias can take many forms, driving human creativity and cultural identities, but just as readily leading to disagreement and conflict. Bias is embedded in our genes—for example, the tendency to taste the bitterness of phenylthiocarbamide (PTC) is linked to a small number of mutations in a bitter taste receptor family gene, and this determines if you taste the bitter compounds in some vegetables. When genes like this are involved, bias—say against broccoli or Brussels sprouts—can be an inescapable feature of our biological make-up. In short, bias is a defining characteristic of people in all walks of life . . . which leads us to a rather big problem: if our biases define who we are, and if you and I are really very different because of them, how do we avoid letting our opinions and beliefs get in the way when trying to identify true causes of biological events?

In this chapter, we will follow the timeline in Figure 5.1 to show how bias has shaped the design of studies seeking to explain the cause, distribution, and control of diseases. From this, we see how attempts to control bias have given rise to one of the most experimentally powerful and life-saving fields in biomedicine: epidemiology.

Figure 5.1 A timeline of events, from recognition of bias as a hinder to scientific progress to the science of modern epidemiology.

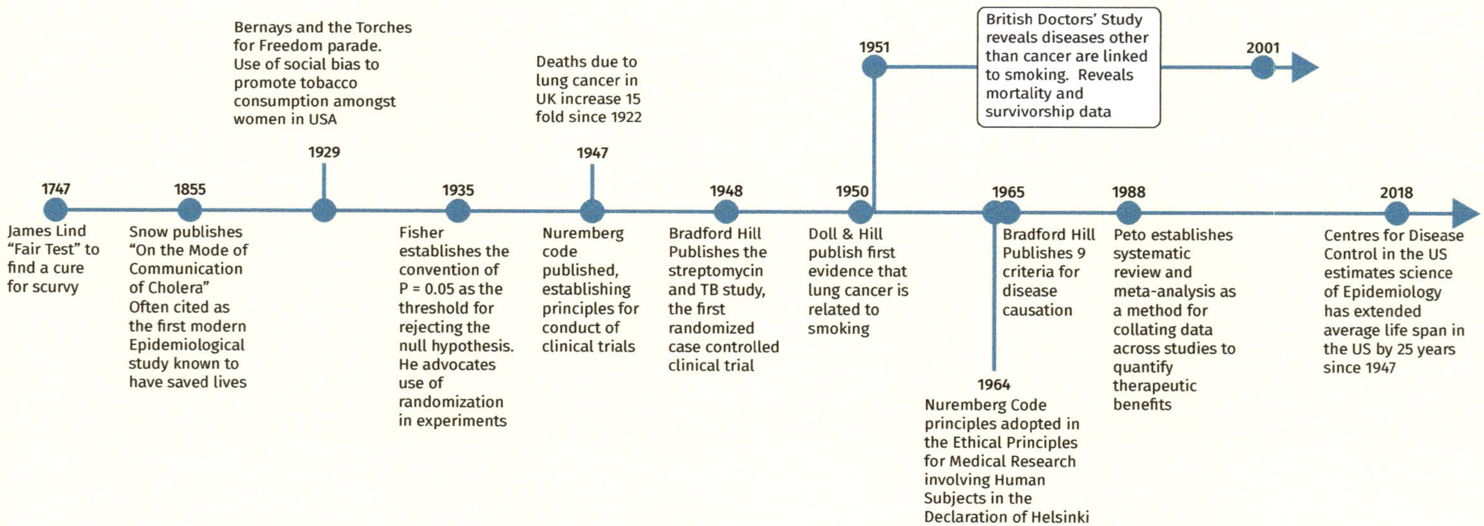

James Lind and scurvy: the first documented clinical trial

Let your mind time-travel back to 1747, the age of naval exploration when the charting of foreign lands by European explorers is in full swing. Most voyages lasted years, and conditions aboard naval vessels were far from comfortable. Crew quarters below decks were poorly ventilated, cramped, and often shared with livestock and vermin—so opportunities for exercise and fresh air were limited. Food consisted mostly of hard-tack biscuits and salted meat; fresh fruit and vegetables were only available at landfall. Disease was rife below decks and chief among these was **scurvy**, a tissue-wasting condition which was the leading cause of death among sailors.

Theories to explain the causes of scurvy were as numerous as the number of doctors who proposed them, with each biased towards their own view of how to treat the condition. These ideas ranged from hard work, putrefaction of the guts because of the ship's diet, bad water, and constipation, to eating salt meat in a damp atmosphere! One ship's surgeon, James Lind, isolated the problem: with so many different ideas, there could be no prospect of finding a cure unless each treatment was tested side-by-side and at the same time.

During a voyage aboard HMS Salisbury in 1747 with the sailors badly affected by scurvy, Lind set up a 'fair test' of the leading treatments of the day in what was to become the first documented clinical trial.

Lind's fair test

James Lind selected 12 sailors with closely matching scurvy symptoms and arranged them into six pairs. They were isolated from the rest of the crew in the ship's fo'c'sle and all were given the same basic diet of water, hard-tack biscuits, gruel, and mutton. Each pair were given one of the test treatments twice per day (Figure 5.2) and the course of the disease was followed over time. There are many similarities between Lind's design and a modern non-randomized before-and-after study (see Figure 5.4). The 6 treatments were:

- one quart of cider per day,
- a gargle of 25 drops of elixir vitriol (sulphuric acid!) three times per day,
- two spoonfuls of vinegar three times per day plus vinegar in all food,
- a half-pint of sea-water to drink once per day,
- two oranges and one lemon per day, and
- an electary (mixed ingredients) of mustard, horse radish, balsam of Peru, and gum myrrh in barley water with tamarind and cream tartare.

What did he find? The sailors who were given oranges and lemons recovered within six days of the start of the treatment. In fact, their recovery was so rapid that they assisted Lind in treating the other sailors who only

Figure 5.2 Citrus fruit—or actually, the vitamin C they contained—proved the solution to scurvy. Amazingly, the most palatable medicine Lind tried was the best! Credit: Anthony Short

showed minor improvements or continued to get worse. Lind had established the first definitive clinical evidence for the use of citrus fruit to treat scurvy—a real breakthrough!

Surely, this was the end of our story and scurvy became a thing of the past? What actually happened next is, to modern eyes, very strange—and provides a valuable lesson about the effect of bias in scientific discovery. The Admiralty commissioned Lind to systematically gather all evidence about the nature, causes, and cure for scurvy which he published in his famous report, '*A Treatise on the Scurvy*' in 1753 (Figure 5.3). In the opening pages, he acknowledges that biased viewpoints have slowed work to find a cure, writing: '*…it is no easy matter to root out old prejudices, or to over-turn opinions which have acquired an establishment by time, custom and great authorities….*'. You might imagine his discovery that citrus cures scurvy would be trumpeted from every page, yet his ground-breaking clinical trial is buried in the middle of the book (p.149).

Although Lind acknowledges several reports of the benefits of citrus and berry juice, he concludes that the nutritional value of the sea-faring diet is adequate and that clean air and exercise (particularly horse riding!) should be recommended to prevent the disease. It seems that Lind's observation is lost to the 'old prejudices and opinions' that he was attempting to avoid. Amazingly it was nearly another half-century before citrus was formally recognized by the medical profession and Admiralty as a cure for scurvy. This is a clear demonstration of the way bias of all kinds acts as a potent force in science and can slow progress by creating blind alleys which some researchers pursue for years.

Figure 5.3 James Lind's work, *A Treatise on the Scurvy*, was published in 1753. It did not have the revolutionary effect on deaths from scurvy that you might expect!

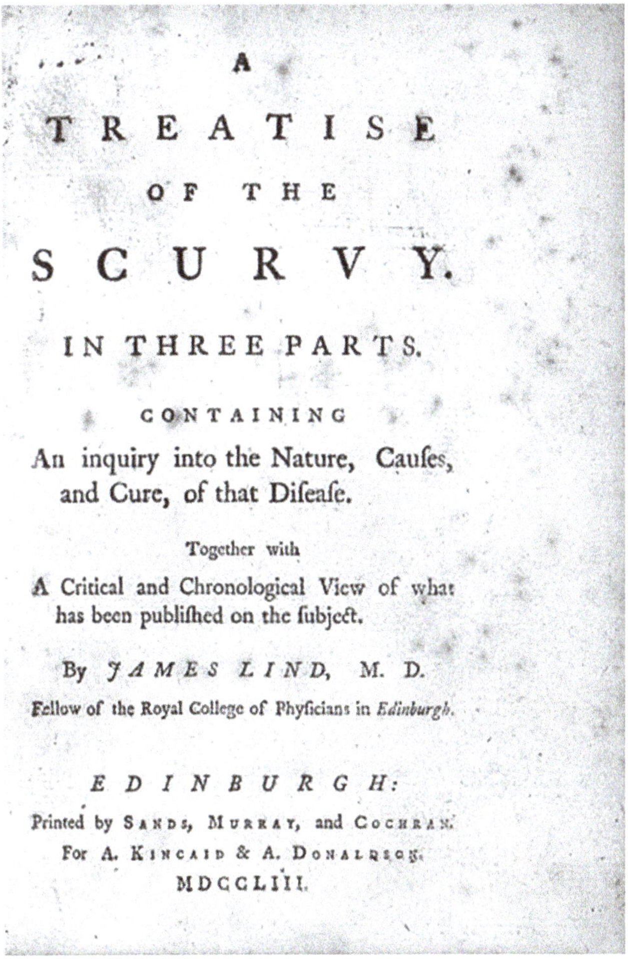

A

TREATISE

OF THE

SCURVY.

IN THREE PARTS.

CONTAINING

An inquiry into the Nature, Cauſes, and Cure, of that Diſeaſe.

Together with

A Critical and Chronological View of what has been publiſhed on the ſubject.

By *JAMES LIND*, M. D.

Fellow of the Royal College of Phyſicians in *Edinburgh*.

EDINBURGH:

Printed by SANDS, MURRAY, and COCHRAN.

For A. KINCAID & A. DONALDSON.

MDCCLIII.

The birth of epidemiology

Although Lind established the principle of 'fair testing' which was to become the central concept of clinical trials in the future, the next breakthrough in the battle against bias in experimentation did not occur for another 100 years. The industrial revolution in the UK brought increased migration of labour from rural communities into cities, where poor sanitation and cramped living conditions created the perfect conditions for the spread of infectious disease. Families relied on water drawn from common-use pumps located at street junctions, and the water supplying these came from rivers that were often treated as open sewers.

From 1832 several cholera outbreaks had occurred across London. Cholera had long been the scourge of battlefields (Figure 5.4), but in the 19th century it took hold of many major cities. At the time people did not

Figure 5.4 Robert Seymour's powerful image of cholera felling soldiers—and everyone in its path—was painted long before anyone understood the spread of this killer disease.

understand the concept of bacterial disease transmission—the germ theory of disease was still being developed. Most people believed cholera was spread by inhalation of the bad air or miasma—imagine sewer smells mixed with smog—that was common on city streets.

A London physician, Dr John Snow, challenged this belief, suggesting that cholera was spread by ingestion (being swallowed), not inhalation (breathing in). In 1854, during a particularly serious cholera outbreak in the Soho area of London, he proved his theory by plotting the occurrence of deaths on a street-map of the area. The pattern showed a clustering of infection around a pump in Broad Street where, tellingly, the incidence of death reduced with distance from the pump (Figure 5.5). Snow identified the water in the pump as the source of the disease and arranged to have the pump handle removed to prevent further infection.

The water for the pump had been taken from a stretch of the Thames that served as an open sewer. Snow's careful investigation showed the link between faecal waste, a water-borne infectious agent (now known to be the bacterium *Vibrio cholerae*), ingestion, and gut infection. Because cholera causes violent diarrhoea, large amounts of contaminated waste were flushed back into the sewer system, further intensifying the outbreak. Snow's investigation was, effectively, a double-blind trial. Neither he nor the subjects he was studying had any knowledge of the causes of disease transmission and so this was a fair, unbiased exploration of the miasma inhalation versus the ingestion theory. Snow's methodical work dispelled the idea of foul air as a cause of cholera. His report of the outbreak, published in 1855, founded the new science of **epidemiology**: the study of the incidence, distribution, and possible control mechanisms for diseases within populations.

Figure 5.5 John Snow's cholera map showed the location of deaths as black tally marks which clustered around the Broad Street pump (left). In this modern image, his data have been used to make a heat map (right) which shows the spread of the epidemic around the pump. Locating the Broad Street pump as the centre of a local cholera outbreak changed the way we look at the spread of infectious diseases forever.

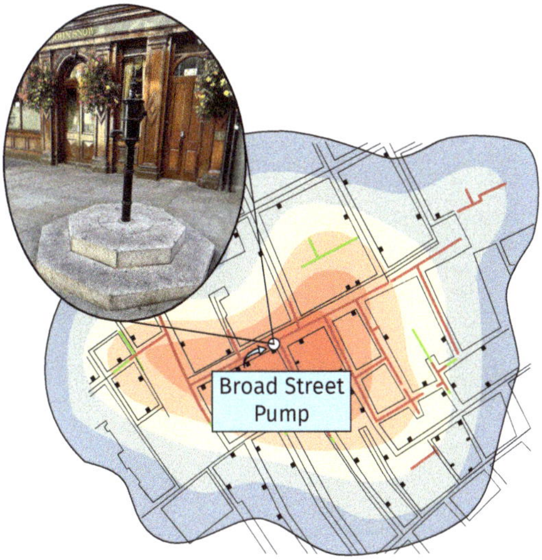

Managing bias

At the turn of the 20th century, as scientists and doctors were getting to grips with how infectious diseases were spread, and how they might be prevented, mathematicians were developing modern concepts of probability, randomness, and chance. As we saw in Chapter 3 and Gossett's role in developing Student's t-test, much of the early understanding of bias and probability came from crop trials with fields of study like medicine tending to be driven by personal guesswork and experience From this emerged the idea that randomness could be used as an approach to tackle experimenter bias and provide a scale against which the likelihood of events could be measured.

Our modern understanding of statistics owes much to Ronald A. Fisher, who was among the first to advocate randomization, hypothesis testing, and use of probability in experimentation. In his book, *The Design of Experiments*, published in 1935, Fisher makes the point that experiments should be unbiased and provide a true measure of the likelihood of encountering the null hypothesis purely by chance. Whilst head of statistics at the agricultural Rothamsted Experimental Station in Harpenden, UK in the 1920s, Fisher and his assistant Yates faced a particular problem. How do you run crop trials in a field where there are so many things that might bias the experimental outcome? For example, drainage across the field might be uneven, exposure to sun might vary during the day, perhaps the distribution of nutrients might be higher in one corner than elsewhere. Fisher and Yates suggested that crop trials should be organized in a grid called a **Latin square** where each treatment would occur once in a different part of the grid in every column and every row (Figure 5.6). A letter from the Latin or Roman alphabet would signify a

Figure 5.6 This is an example of a Latin square experiment (A) which tests the effects of fertilizer and water conditions on carrot growth. Each condition is given a Latin letter symbol which appears once in each row and column of the square for each experimental variable. The photograph (B) shows how a Latin square crop trial appears from above and it is easy to see how the different treatments in each square have influenced plant growth.

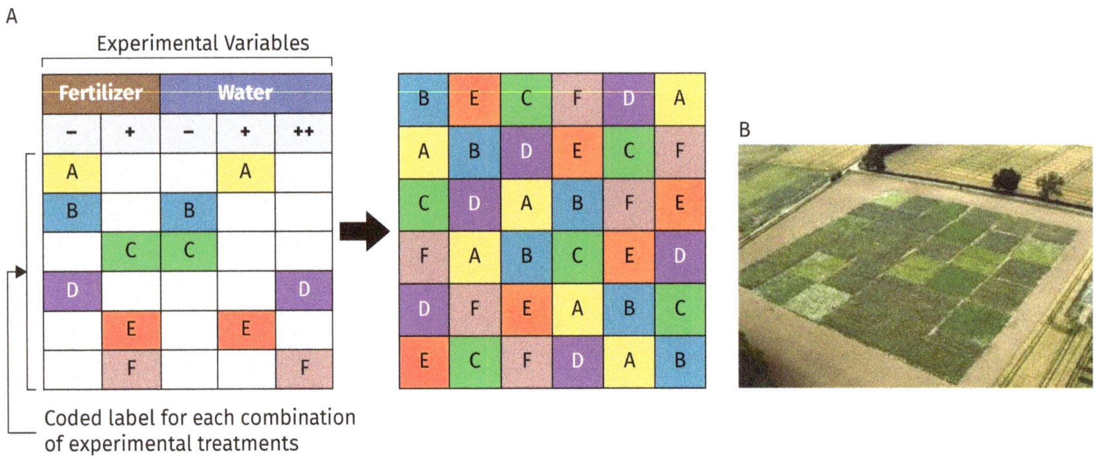

particular treatment, hence the name. This would mean that underlying variation in things like soil quality or sun exposure would be controlled across the experiment. In allowing the measurement of three or more factors, the results from Latin square experiments can be analysed using ANOVA in a way which minimizes the effects of bias and uncontrolled factors. Today, the Latin square approach is used in many areas of scientific research, including medicine, to control for hidden causes of variance in an experiment.

Applying principles of randomization to clinical trials

Wars produce ideal conditions for disease to spread by forcing people to migrate, gather in poor living conditions, starve, manage without proper sanitation—and often, all in the absence of proper medical care. After World War I, Spanish Flu claimed more lives than had been lost during the war, and tuberculosis (TB), a bacterial disease which had been in decline before the war, spread rapidly through Europe, just as nations were rebuilding their infrastructure and economies (Figure 5.7).

At this time, TB had life-changing consequences for a newly qualified Royal Navy Air Service pilot called Austin (Tony) Bradford-Hill. Son of a medical physiologist, Hill contracted tuberculosis in 1917 whilst travelling through Europe to his first posting in Greece. On his return home he was exposed to the grim treatment of the day for TB: nine months of bed rest with surgically controlled lung collapse and reinflation (artificial pneumothorax) given at periodic intervals to 'rest the lung'. At a time when the prognosis for this disease was poor, Hill cheated the likely outcome of death and managed to recover. Under the mentorship of Major Greenwood, friend and Senior Statistical Officer at the newly formed Ministry of Health, he learned to apply his interests in mathematics and social science to the causes and treatments for disease. Hill became the

Figure 5.7 Graph showing incidence of TB from 1885 to 1935—the effect of World War I is clearly visible as a peak in the data from 1914 into the post-war period of the 1920s.

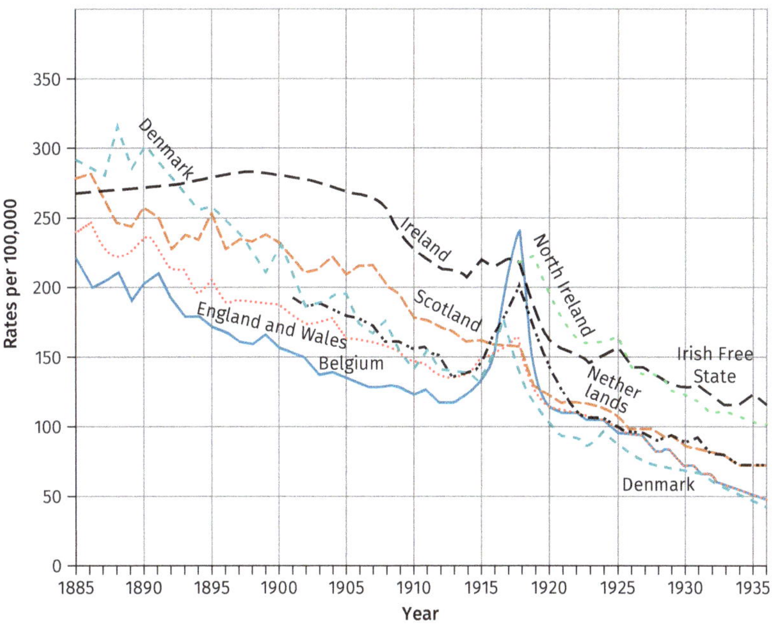

first Professor of Medical Statistics at the London School of Hygiene and, in a highly personal turn of events, was appointed Chief Statistician by the Medical Research Council to determine if streptomycin, a new antibiotic, was effective in treating patients with TB.

Hill knew Fisher well and recognized the danger of 'systemic bias' in clinical studies. He reasoned that:

- Researchers have an interest in the outcome of their experiments which favours positive results over neutral or negative findings.
- Patients have a vested interest in being cured and so may feel particularly motivated to report positive outcomes.
- Institutions have a vested interest in their clinicians making patients better and so will favour research (and researchers) that make patients better.

He addressed these problems by designing a randomized case-controlled trial of streptomycin as a treatment for TB (see Figure 5.8) which blinded experimenters in four different ways:

1. Just as James Lind had done, 200 years before, patients were assessed to ensure similar disease characteristics and severity. The study ran at several hospitals across the UK and so took account of local treatment preference and environmental influences. A statistician provided a random sequence code for control + bed rest (C case) and experimental + bed rest (S case) treatments and the key to each code was sealed in an envelope until the end of the experiment.

2. Patients were randomly assigned to a treatment code by an independent treatment allocation panel.
3. The coded treatment was administered by an independent medical officer with no involvement in data collection or analysis.
4. Data was collected from each patient and analysed by an independent data analysis panel.

At first, this idea was met with dismay. How could a potentially life-saving treatment be denied, at random, to half the patients in the trial? What if more patients received one treatment than the other? Why do so many different separate groups of researchers need to be involved? Hill's response was that there was no indication, at that time, that streptomycin would have any benefit. In fact, it could conceivably make patients worse! There was therefore an even risk of cost/benefit across the treatment groups. He also argued that if the study was large enough then the number of control and experimental groups would balance out. To make this work, several independent groups of experimenters would have to be involved!

Figure 5.8 The structure of Hill's randomized controlled clinical trials—a revolutionary experimental design idea which is now commonly used in clinical trials.

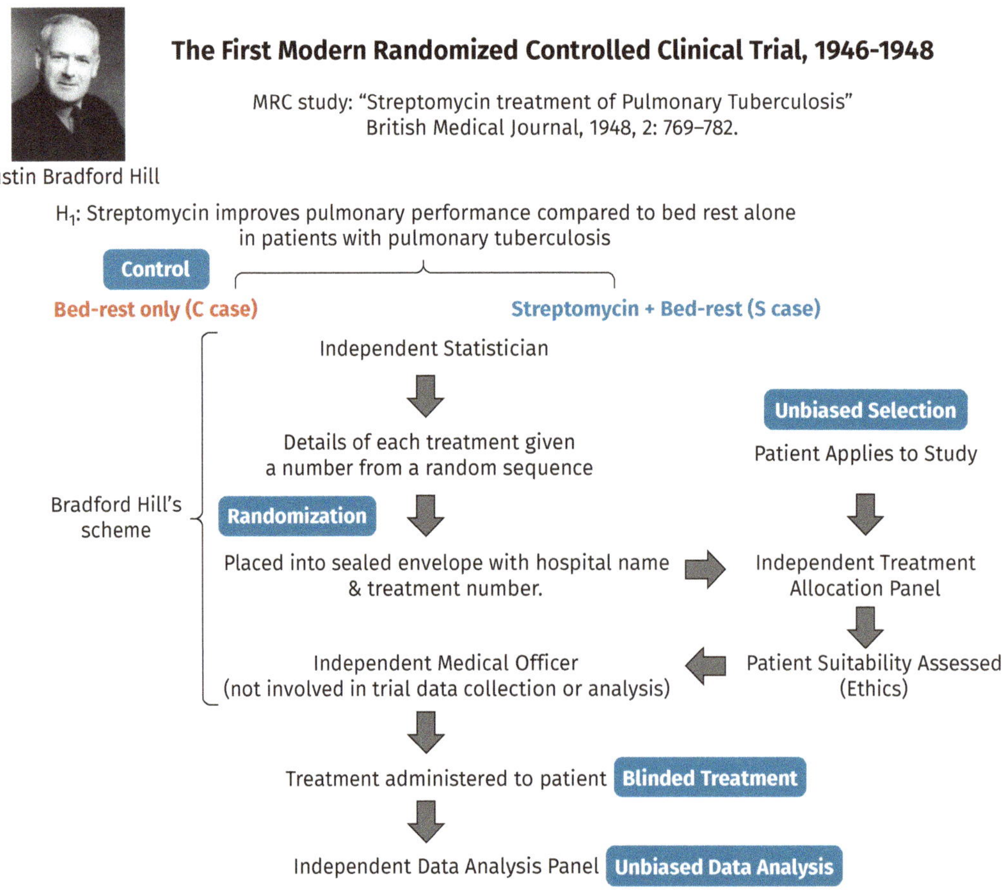

Table 5.1 Results of Hill's streptomycin Randomized Controlled Trial, comparing the appearance of the lungs of the patients before and after treatment.

Radiological Assessment	Streptomycin Group		Control Group	
Assessment of radiological appearance at 6 months compared to appearance on admission	**Number of patients**	**Percentage of the total (%)**	**Number of patients**	**Percentage of the total (%)**
Considerable improvement	28	51	4	8
Moderate or slight improvement	10	18	13	25
No change	2	4	3	6
Moderate or slight deterioration	5	9	12	23
Considerable deterioration	6	11	6	11
Death	4	7	14	27
Total	55	100%	52	100%

The actual results from this randomized controlled study, the first of its kind, are shown in Table 5.1. The greatest difference is in the number of patients who showed considerable improvement (28 in the streptomycin group; four in the control group), and also the number of deaths (four in the streptomycin group and 14 in the control group). The numbers of patients in each group are almost even, supporting Hill's view that if the sample size is large enough then the distribution of patients across each group will even out. Streptomycin was used as the first-line treatment for tuberculosis and served as a model to develop the next generation of antibiotics for bacterial diseases. Just as importantly, Hill's randomized controlled trial (RCT) design became the essential format for clinical trials of this kind to the present day.

Bigger picture 5.1

Hill's Criteria for Disease

Hill's work has had far-reaching impacts on medicine. Following his ground-breaking work on tuberculosis, Hill went on to establish a framework for identifying true causes of disease. In a speech to the Royal Society of Medicine in 1965, he identified nine criteria which should be met when testing associations between environmental factors (for example, viruses, bacteria, pollution, toxins) and disease (Figure A). Hill's causal criteria represented an important step forward in limiting sources of bias and directing research along a rational line of thinking, and are perhaps now more relevant than ever as techniques for scientific investigation become more powerful.

Figure A Diagram showing Bradford-Hill's criteria for disease causation.

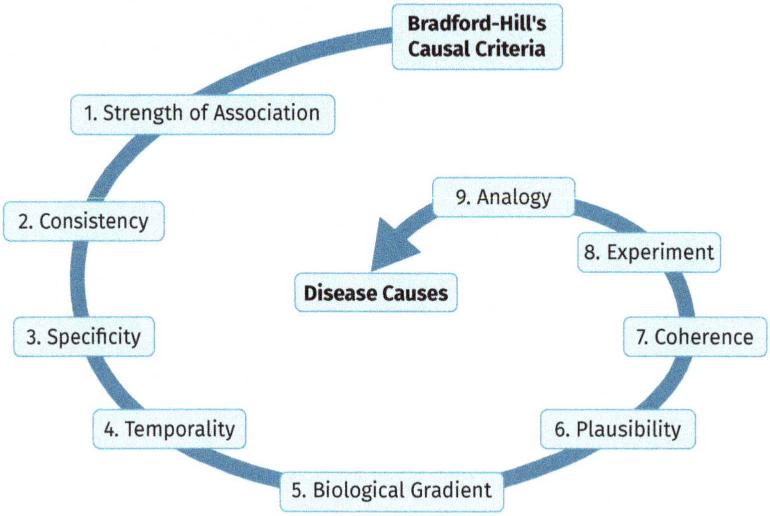

1. **Strength of association**: It is often true that multiple factors drive diseases, so it is really important to discover the strength of the association between a particular factor and a given disease. For example, asthma is often associated with inhaled irritants like pollen or dust, but its severity is also linked to consumption of fat-rich foods and whether or not a person has the skin condition, eczema. Which is more important when it comes to treating the disease? This has led to the development of statistical techniques to find out which, among several possible factors, has the greatest probable association with the disease.
2. **Consistency**: It should be possible to observe strongly associated disease-causing factors at different geographical locations using separate study populations and types of measurement. Hill was particularly aware that individual studies carry bias risk and so observation across multiple studies is the only way to minimize this problem.
3. **Specificity**: A single, defined mechanism has a consistently high strength of association with the disease. This could be a single environmental factor (for example duration of ultraviolet irradiation and skin cancer (melanoma)) or a biological process (e.g. mutations in a gene which result in the production of a dysfunctional protein, for example the deletion of the phenylalanine codon in the *CFTR* gene which disrupts chloride ion movement through the CFTR ion channel seen in cystic fibrosis).
4. **Temporality**: Exposure to a specific disease-causing factor should come before the disease develops and progress predictably over time. Think, for example, how bacteria in out-of-date processed meats might cause food poisoning symptoms which develop, then clear over time.
5. **Biological gradient**: Simply put, the greater the exposure to the disease-causing agent, the greater the risk of developing the disease. Nowadays, much emphasis has been placed on identifying the 'threshold' levels of exposure to a suspected causal factor which can be achieved

without causing the disease to develop (known as the 'no-observable-adverse-effect level' or 'NOAEL'). This enables scientists to study how cell processes protect the body at this level of exposure and what goes wrong when the threshold is exceeded.

6. **Plausibility**: The logical causal link between biological factors which drive the transition from health to disease. With the development of new bioinformatic tools to identify plausible causes of disease, this is becoming ever more complex and has led to the concept of 'precision medicine', a discipline whose goal is to establish detailed plausible links between biological processes and disease for the development of treatments capable of targeting molecular mechanisms around the disease threshold.
7. **Coherence**: Cause and effect should fit with current understanding of the disease mechanism and will evolve in parallel as further discoveries are made and with the development of new biological tools.
8. **Experiment:** Experimental research often involves manipulation of single factors which plausibly fit a disease mechanism. The challenge for scientists in the future is to map how multiple factors interact with each other to cause particular forms of disease.
9. **Analogy**: When one causal agent is known, the standard for accepting a second causal criterion which shares similarities with the first is lowered since a crucial frame of reference has been established.

As with many central concepts in science, the 'Hill Criteria for Disease Causation' began as a discussion framework for streamlining ideas in epidemiology, but they have become one of the most valuable tools for identifying factors which cause disease. With the advent of methods for analysing large and complex data sets, the criteria of specificity, biological gradient, plausibility, coherence, and experiment have become vital guiding concepts in eliminating bias and identifying true causes of disease.

? Pause for thought

- How might you explore and explain causality of the common cold using Bradford-Hill's Causal Criteria?

Linking lung cancer to tobacco smoking

While Fisher and Hill were developing their ideas about randomization and experimental bias constraint, a social revolution was happening in America which was to test the strength of these concepts to the full. The suffragette movement at the turn of the 20th century led to dramatic social changes, with women gaining the right to vote, enter government, and seek equal opportunity in the workplace. Up to this time, for a woman to be seen smoking in public was considered taboo and immoral. During World War I many women

Figure 5.9 Cynical advertising in the 1920s attempted to persuade women that smoking was both emancipated and would help them stay slim!

had performed jobs previously done by the men away fighting the war, and by the 1920s, cigarettes became feminist symbols of social freedom, equality, and the right to self-determination. This new market potential was not lost on cigarette companies. The American Tobacco Company embarked on an aggressive advertising programme led by a public relations expert called Edward Bernays. The aim was to establish cigarettes as the symbol of feminist emancipation by promoting a sophisticated image of the female smoker as an upwardly mobile defender of free choice and of equal gender rights. The advertising campaign targeted young female smokers by promoting the activity as a healthy alternative to sweets to aid weight loss (Figure 5.9).

In 1929, Bernays paid a group of women to march in New York's Easter Sunday Parade carrying banners reading 'Torches for Freedom'. and smoking cigarettes he had supplied. With support from feminist activists, who encouraged women to join in, the event was photographed widely and the images spread around the world, leaving a lasting impact on how tobacco companis marketed cigarettes to women. Before the campaign, women purchased around 5% of cigarettes sold—this rose to 33% by 1965. Subtle targeting of different groups through carefully calculated imagery remains an active and effective form of tobacco advertising around the world even now, in countries where tobacco advertising is not already banned on health grounds!

The rise in tobacco smoking created a health crisis time bomb. In England and Wales, the number of deaths from lung cancer rose fifteen-fold from 1922 to 1947 (Figure 5.10), but few people accepted—or wanted to accept—smoking as the cause. In an echo of Snow's search for the cause of cholera,

Figure 5.10 By 1945, deaths from lung cancer had risen to more than 15 times the number recorded in 1900. This change coincided with a rise in cigarette smoking linked to advertising campaigns such as Bernay's Torches for Freedom Parade in 1929.

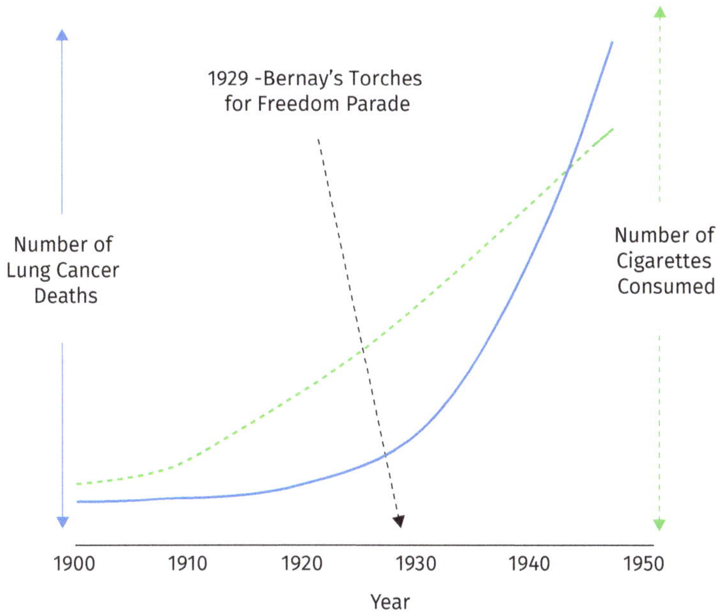

many blamed poor air quality from industrial pollutants and motor car exhausts. How were scientists to differentiate between these two plausible causes of lung cancer without succumbing to strongly held, biased, public opinions? Following the success of the streptomycin and tuberculosis trial, Hill enlisted the help of a young doctor called Richard Doll who had been de-mobilized from the army at the end of World War II. Doll had met Hill whilst completing a course in medical statistics at the London School of Hygiene, and both quickly recognized a shared interest in applying mathematics to medical problems. They embarked on two studies to test the hypothesis that tobacco smoking causes lung cancer. The first was an observational case control study which compared the incidence of smoking among lung cancer and non-cancer patients (see Case Study 5.1).

Although this study was comparatively small, it provided the first solid evidence that smoking might be linked causally to lung cancer—but you notice the care that Doll and Hill took not to bias their interpretation by over-stating the link in their summary:

> 'It is not reasonable, in our view, to attribute the results to any special selection of cases or to bias in recording. In other words, it must be concluded that there is a real association between carcinoma of the lung and smoking . . . This is not necessarily to say that smoking causes carcinoma of the lung. The association would occur if carcinoma of the lung caused people to smoke or if both attributes were end-effects of a common cause. The habit of smoking was, however, invariably formed before the onset of the disease . . . so that the disease cannot be held

to have caused the habit; nor can we ourselves envisage any common cause likely to lead both to the development of the habit and to the development of the disease 20 to 50 years later. We therefore conclude that smoking is a factor, and an important factor, in the production of carcinoma of the lung.'

Scientific approach 5.1

Proving the link between cancer and tobacco smoking—χ^2 analysis of a case-controlled study

Doll and Hill's first study, published in 1950, used a retrospective observational case control design which, in their words was '... to be carried out on a sufficiently large scale to determine whether patients with carcinoma of the lung differed materially from other persons in respect of their smoking habits or in some other way which might be related to the atmospheric pollution theory' (Figure A).

Figure A We now know that tobacco smoke contains many different carcinogens and causes an increased risk of many cancers, heart disease, and other health issues. It all began with the work of Hill and Doll establishing a link between smoking and lung cancer.

One might imagine that the obvious thing to do would be to compare the incidence of lung cancer between smokers and non-smokers but this design would presuppose the hypothesis to be true; in other words, it would be biased towards the idea that smoking does indeed cause lung cancer. Doll and Hill's study posed the question in a slightly different way. They interviewed a cross-section of patients, carefully selected to be matched for age and sex, who were attending hospital for different medical reasons. Those with lung, kidney, colon, or rectal cancer were categorized as 'test subjects' and those attending for other medical reasons were categorized as 'control subjects'. A smoking history was taken from both groups after subjects had been recruited to both groups, and only then was the health of the smokers considered. By setting up the study in this way, Doll and Hill asked the question in a very precise way: If you have lung cancer, are you also more likely to be a smoker? With this question, their study captured three key elements:

1) avoidance of biased recruitment of smokers into the test group
2) the risks of cancer for men and women could be measured (particularly important given the active marketing of cigarettes to women by Bernays and the American Tobacco Company)
3) links between smoking intensity and any disease which had caused the patient to attend hospital (see Hill's biological gradient in Case Study 5.1).

Doll and Hill argued that if smoking had no link to lung cancer, the test and control groups would show the same results. They tested this idea using Pearson's **Chi squared (χ^2) test** which compares the distribution of the test to control group data (see Chapter 2 for explanation of data distributions).

The χ^2 formula is Equation 5.1:

$$\chi^2 = \sum\left(\frac{(\text{Observed} - \text{Expected})^2}{\text{Expected}}\right)$$

The observed values are measured from test and control group patients. These are used together to determine the range of values that would be expected if smoking had no effect (don't confuse control group data for 'Expected values', a common error!). Table A contains data from Doll and Hill's original paper (*British Medical Journal*, 1950) which shows the relationship between number of cigarettes smoked up to the onset of disease and incidence of either lung cancer or non-cancerous disease for men and women.

It is not easy to see patterns from the data in the table, but Hill believed that it should be possible to talk the reader through the logic of any statistical test. Let's take a stepwise approach using the male data (top two data rows) to explore how χ^2 can be used to detect patterns. Follow this example through and then try to reproduce the χ^2 values for females, noting that the number of women smoking 15 or more cigarettes per day have been grouped together. You might find your own answers differ a little bit depending on how you round your numbers but the process you follow will be correct.

Table A Number of Cancer Patients Categorized According to Smoking Behaviour.

	Number of Cigarettes Smoked Daily					Probability Test
Disease Group	**1**	**5**	**15**	**25**	**50**	
Males: Lung-carcinoma patients (647)	33 (5.1%)	250 (38.6%)	196 (30.3%)	136 (21.9%)	32 (5.0%)	χ^2 = 36.95; n=4; P<0.001
Control patients with diseases other than cancer (622)	55 (8.8%)	293 (47.1%)	190 (30.5%)	71 (11.4%)	13 (2.1%)	
Females: Lung-carcinoma patients (41)	7 (17.1%)	19 (46.3%)	9 (22.0%)	6 (14.6%)	0 (0.0%)	χ^2 = 5.72; n=2; 0.05<P<0.10
Control patients with diseases other than cancer (28)	12 (42.9%)	10 (35.7%)	6 (21.4%)	0 (0.0%)	0 (0.0%)	

Table V from Doll and Hill (1950) showing rise in death rate from cancer of the lung and consumption of tobacco

Step 1. Calculate expected values. The experiment gives a snapshot of the distribution of illness in the population and so the total numbers of smokers in both lung carcinoma and control groups can be used to calculate the overall expected distribution of illness. Using data from males we derive Table B.

The expected values are determined from the proportion of all patients falling into each smoking category.

- Smokers with lung cancer make up 647/1269, or 51%, of the total population.
- Smokers with other diseases make up 622/1269, or 49% of the total population.

We can use these proportions to calculate the expected values for both groups using the formula:

$$\text{Expected} = \text{total patients in a smoking category} * \left(\frac{\text{Total patients in the test group}}{\text{Total patients in the study}}\right)$$

Table B Incidence of Cancer and Daily Cigarette Consumption (Males).

	Number of Cigarettes Consumed Per Day (Males)					
Test Group	**1**	**5**	**15**	**25**	**50**	**TOTAL**
Observed Test: Lung carcinoma	33	250	196	136	32	**647**
Observed Control: Patients with diseases other than cancer	55	293	190	71	13	**622**
SUM	**88**	**543**	**386**	**207**	**45**	**1269**

Calculating the total number of male patients in each observed smoking category, the number in each test group, and the total number in the study.

Table C Observed and Expected Numbers for Different Cancers and Cigarette Consumption Per Day (Males).

	Number of Cigarettes Consumed Per Day (Males)				
Test Group	**1**	**5**	**15**	**25**	**50**
Observed Test: Lung carcinoma	33	250	196	136	32
Expected Test: Lung carcinoma	45	277	197	106	23
Observed Control: Patients with diseases other than cancer	55	293	190	71	13
Expected Control: Patients with diseases other than cancer	43	266	189	101	22

Calculating the expected number of male patients in each smoking category.

For the first smoking category of 1 cigarette per day, the expected results would be:

- Expected values for lung carcinoma: 88 × (647/1269) = 45 patients
- Expected values for other diseases: 88 × (622/1269) = 43 patients

This is repeated for each smoking category to create a table of observed and expected values (Table C). To make things easier, round your numbers to the nearest whole number.

Step 2: Determine the critical value of χ^2.

Once we have found the observed and expected values, the χ^2 formula is applied to each set of test data and a χ^2 value is calculated by adding up all the values in the table.

$$\chi^2 = \sum\left(\frac{(\text{Observed} - \text{Expected})^2}{\text{Expected}}\right)$$

Doll and Hill rounded their expected results to 2 decimal places so to get their exact result we would have to do the same. So: for people with lung cancer who smoke just 1 cigarette a day the calculation is:

$$\left(\frac{(\text{Observed} - \text{Expected})^2}{\text{Expected}}\right) = \left(\frac{(33 - 44.87)^2}{44.87}\right) = \left(\frac{-11.87^2}{44.87}\right) = \left(\frac{140.90}{44.87}\right) = 3.14$$

For people with other diseases smoking just 1 cigarette a day:

$$\left(\frac{(55 - 43.13)^2}{43.13}\right) = \left(\frac{11.87^2}{43.13}\right) = \left(\frac{140.90}{43.13}\right) = 3.27$$

Results for all the observations are in Table D.

To find chi squared (χ^2), sum all the results in your table:

$$\chi^2 = 3.14 + 3.27 + 2.60 + 2.71 + 0.003 + 0.003 + 8.79 + 9.14 + 3.58 + 3.72$$
$$= 36.96$$

Table D Data to be Summed for the χ^2 Calculation (Males).

	Number of Cigarettes Consumed Per Day (Males)				
	1	**5**	**15**	**25**	**50**
Test: Lung carcinoma	3.14	2.60	0.003	8.79	3.58
Control: Patients with diseases other than cancer	3.27	2.71	0.003	9.14	3.72

Step 3: Determine number of degrees of freedom.

This is determined as:

$$
\begin{aligned}
\text{degrees of freedom} &= \text{number of test conditions(cols.)} - 1 \\
&\quad * \text{number of test categories(test and control} - 1) \\
&= (5 - 1) * (2 - 1) \\
&= 4 \text{degrees of freedom}
\end{aligned}
$$

Step 4: Determine probability that the critical value of χ^2 matches the distribution of the test data.

Our results agree with those of Doll and Hill, where $\chi^2 = 36.95$, $n = 4$.

Figure B shows a section of the χ^2 distribution tables and you will notice that χ^2 values become bigger as the probability threshold, α, becomes smaller. At $\alpha = 0.05$ and 4 degrees of freedom, the critical value of χ^2 is 9.488. Our calculated χ^2 value of 36.96 is much greater than this value and so the null hypothesis can be rejected. We can conclude that there is an association between smoking and lung cancer in males.

Doll and Hill went on to use this method to examine several different measures of the link between smoking and lung cancer (for example age at which people start to smoke, total tobacco consumption, peak life-time smoking activity) and in each case, the χ^2 test revealed a strong association between tobacco and lung cancer. Importantly, they also demonstrated that stopping smoking reduces the chances of developing lung cancer and so they provided the first statistical, unbiased evidence to counter the argument from tobacco companies, and advanced by Bernays, that smoking is good for you.

Figure B Calculating the total number of male patients in each observed smoking category, the number in each test group, and the total number in the study.

Degrees of freedom	Probability ($\geq X^2$)			
	0.1	**0.05**	**0.025**	**0.01**
1	2.706	3.[illegible]1	5.024	6.635
2	4.605	5.[illegible]1	7.378	9.21
3	6.251	7.[illegible]5	9.348	11.345
4		9.488	11.143	13.277
5	9.236	11.07	12.833	15.086

Pause for thought

Doll and Hill's study attempted to capture how factors like sex, socio-economic background, location, intensity of smoking, and type of tobacco smoked might influence risk of lung cancer.

- How might a Latin square help to control the influence of these things on the results—and what are the limitations of such an approach when looking at experiments involving people?

Studies over time: building the evidence

The first study carried out by Doll and Hill, in Case Study 5.1, was relatively small. Richard Doll went on to lead a much larger **longitudinal study** into the effects of smoking. This involved 40,000 participants and was to last for 50 years! Doll recruited only doctors into this study, believing their professional interest in the results would encourage reliable participation over such a long study period. It became known as the **British Doctors Study** and ran from 1951 to 2001. It confirmed the link between smoking and cancer in Doll and Hill's first study but its large size and duration revealed several key findings.

1. An accurate projection of life expectancy among smokers became possible (Figure 5.11).
2. It demonstrated that giving up smoking at any age improves life expectancy compared to life-long smoking.
3. It revealed associations between smoking and vascular diseases that cause heart disease, angina, and stroke which Doll and Hill's 1950 study had been too small to see.

Figure 5.11 The British Doctors Study led by Sir Richard Doll shows how smoking affects life expectancy—and the effect of giving up the habit.

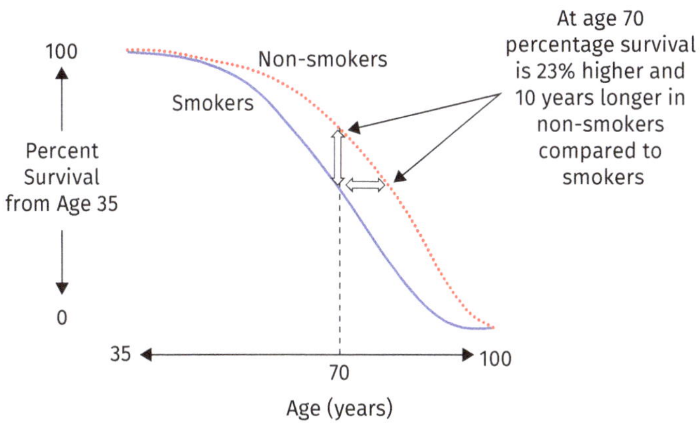

The British Doctors Study demonstrated clearly that the regular smoking of cigarettes over a long period is very dangerous but, importantly for public health organizations and governments, there are clear health benefits to giving up at any stage of life.

The sum of all knowledge: systematic review and meta-analysis

By virtue of its large size and duration, the British Doctors Study revealed links between smoking and diseases other than lung cancer which Doll and Hill's much smaller case-controlled study did not find. This illustrates that study size really matters—the greater the size, the better the statistical likelihood of identifying effects, even if very small (see discussion of statistical power in Case Study 5.1). But however good—or large—an individual study, it will have limitations. Increasingly scientists look at the evidence from many studies, using **systematic review** and **meta-analysis** to identify the causes of disease.

A systematic review is a process where researchers gather all of the published literature on a topic then assess it for reliability by looking at things like the design of the study, possible causes of bias, ethics, and conflict of interest. The aim is to produce the most reliable results possible so that anyone with an interest in those data, like doctors, health providers, policy makers, and even general members of the public, can make informed choices and decisions.

To avoid potential bias introduced by the literature searching process, systematic reviews follow a three-step process which declares how the articles were found, screened, and included in the final analysis. This process follows guidelines known as the **Preferred Reporting Items from Systematic reviews and Meta-Analyses (PRISMA)** so that all systematic reviews follow the same procedures and standards. Figure 5.12 shows a PRISMA flow diagram from a systematic review of robotic surgery procedures for treating a disease of the colon. The search terms included 'robotic resection', 'minimally invasive surgery', 'diverticular disease', 'colonic diverticulosis' and 'diverticulitis' using literature search databases MEDLINE, EMBASE, and PubMed.

Part of the PRISMA screening process involves assessing each paper for its eligibility to be included in the final analysis. Each article is assessed for different causes of bias and an overall assessment of bias risk is generated for the systematic review (Table 5.2).

Bias risk is reported using an **eligibility matrix** where each paper is assessed against different causes of bias using a symbol to indicate low, unclear, or high risk of bias in that category. The scores are added together to create an overall assessment of bias risk, like this one from a systematic review of spinal surgery techniques (Figure 5.13).

When papers have been screened for quality and relevance using the systematic review process, the data within them may be combined in a powerful statistical process called **meta-analysis**.

Figure 5.12 PRISMA reporting of a systematic review of robotic surgery outcomes to treat diverticular disease of the gut.

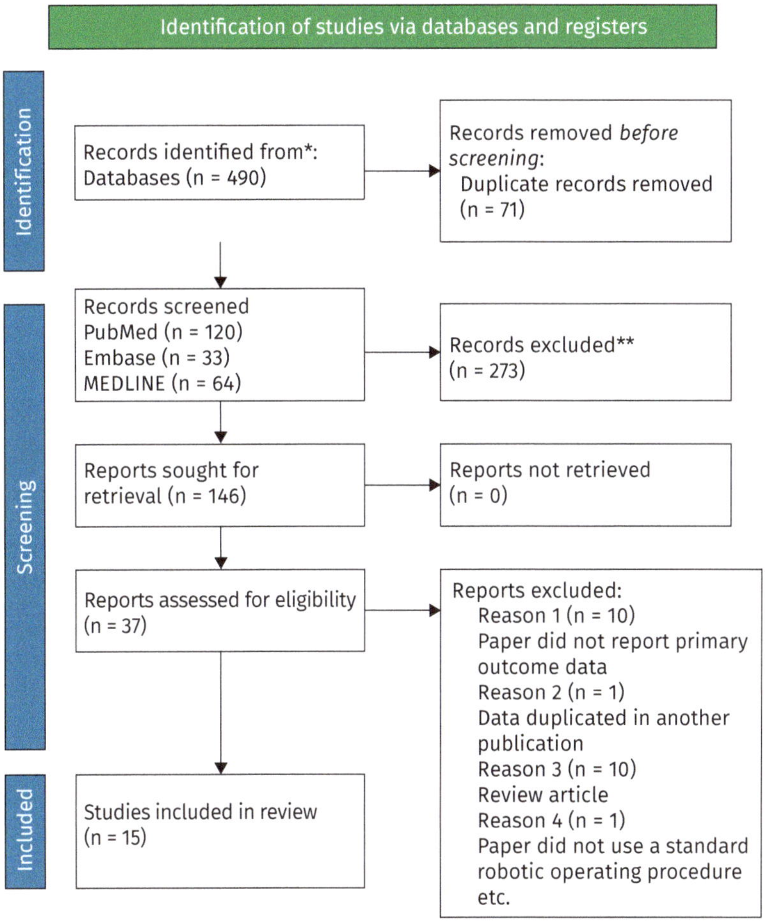

It works by summing data from several studies examining the same question to produce a single, high-resolution measure of an effect. This powerful meta-analysis enabled epidemiologists to make statistical sense of data from studies whose results, if viewed individually, might appear to disagree. Importantly, this approach also eliminated bias caused, for example, by the tendency to favour promising studies over others.

This statistical method was developed by Richard Peto, who collaborated with Doll on the British Doctors Study. Their work had revealed that vascular disease drives the increased occurrence of heart attack and stroke among smokers. The search was on to find inexpensive, safe, and readily available drugs which might reduce risk of these life-threatening events. It had long been known that aspirin was an effective anti-platelet coagulation factor which could thin the blood and reduce the risk of clot formation in vessels supplying the heart, lungs, and brain. However, individual case-controlled studies seemed to show no benefit when it was used in heart attack and

Table 5.2 Categories of bias which are reported in systematic reviews.

Type of Bias	Meaning
Selection Bias • Random sequence generation • Allocation concealment	Participants have been allocated to treatment groups using a randomization procedure, often involving a random sequence of numbers. The sorting of participants into treatment groups is concealed from the researchers, participants, and data analysts. This often requires an independent statistician to assign a code to the allocated groups which is only revealed at the end of the study.
Performance Bias • Blinding of participants and personnel	Like allocation concealment, this involves concealing the placebo or test treatments from the participant and investigators.
Detection Bias • Blinding of outcome assessment	Concealment of the most important measurements from participants and investigators.
Attrition Bias • Incomplete outcome data	Loss of continuity by, for example, patients leaving the study.
Reporting Bias • Selective reporting	Unexplained focus on a particular outcome where the data might support an alternative explanation.

stroke patients. Peto formed a group of scientists called the Antiplatelet Trialists Collaboration and systematically collected data from 25 studies which had tested the hypothesis:

> H_1: Aspirin—either alone or in combination with other anti-coagulants—improves survival among patients who had suffered an incidence of heart attack or stroke.

The first problem was that none of the studies looked for effects of aspirin in the same way. Some tested aspirin alone, whereas others explored aspirin in combination with different drugs. Rather than using absolute result values which could not easily be compared across trials, Peto standardized the results within each study by determining the difference between each treatment and its own control. This was used to calculate an Odds Ratio for the active treatment (see Chapter 2) and then a statistical test was performed on all of the studies lumped together to see if there was any effect.

The second problem was the absence of a consistent experimental approach between papers. So, for example, the studies differed from one another in the number and age of the subjects, their health characteristics, the types of outcomes that were measured and even by the country and continent in which the research was done! This variety meant that bias could be introduced because of different strengths of study design. Peto measured these differences as '**inter-study heterogeneity**' using a χ^2 test to compare each study to the overall expected result and found no significant effect of heterogeneity between studies.

Figure 5.13 Eligibility matrix and assessment of the risk of bias for meta-analysis studies identified using the PRISMA systematic review method.

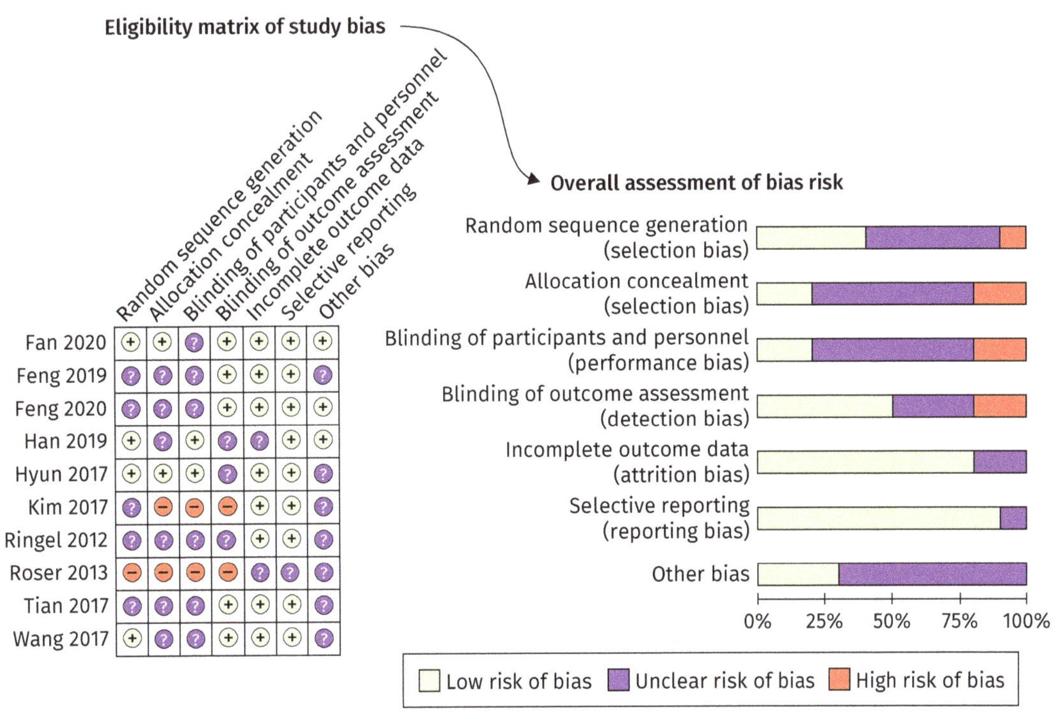

Peto's meta-analysis showed that deaths from vascular disease could be reduced by around 15% using aspirin, with a reduction in non-fatal stroke and heart attack by 30%. These effects may seem modest, but when applied to the population of a medium-sized country such as the UK, they translate into thousands of people whose lives could be extended through aspirin and its anti-blood clotting properties.

Case study 5.1

Forest plots, study power, and heterogeneity

Epidemiologists often summarize meta-analysis results using a **forest plot**. This allows the viewer to compare individual studies to the results of all similar studies together. Peto worked with a group of researchers called the Antiplatelet Trialists Collaboration to develop the forest plot reporting method. Use the following guide to see how to interpret an example forest plot (Figure A).

Figure A An example of a forest plot. Key features of the plot are indicated by the numbers in the green circles. See numbered text for details.

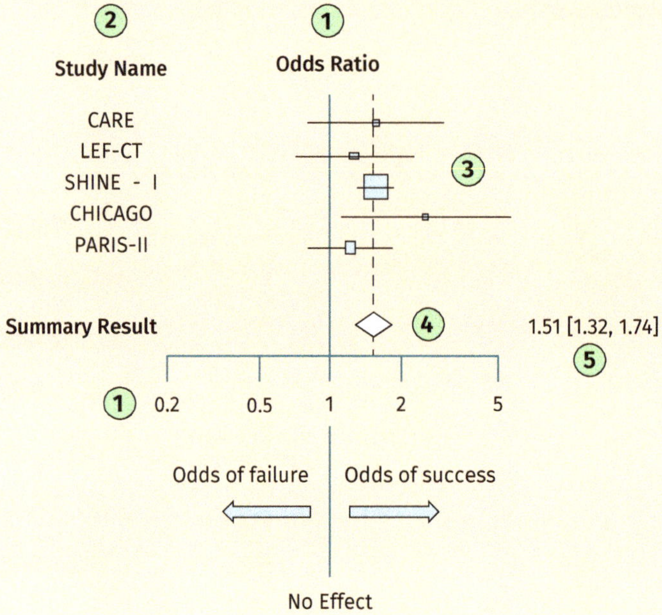

Use the following to help you understand the key features of a forest plot.

1. The odds ratio is given on the x axis and at the top of the forest plot. Values equal to 1 indicate no effect and so a line is projected vertically through the data to show this location on the plot. Interpretation of the odds ratio numbers will depend on the comparison to be made by a meta-analysis. In this example, numbers bigger than 1 suggest a successful outcome and those lower than 1 suggest an unsuccessful outcome.
2. Clinical trials are often given memorable names so that scientists can easily refer to them. Five clinical studies are listed in this example but it would not be unusual for a forest plot to include many more trials.
3. The odds ratio for each study is represented by a square with 95% confidence intervals (CIs) extending horizontally as whiskers on either side. Where CIs intersect the solid line of no effect (Odds ratio = 1) then the null hypothesis cannot be rejected for that study. The size of the square symbol indicates the weight of information which contributes to the overall summary result.
4. The white diamond represents the overall summary result, calculated by combining the size and confidence interval (CI) of each individual study. The centre of the diamond indicates the pooled (aggregate) odds ratio, while its horizontal tips mark the combined 95% CI. In this example, three of the five studies are inconclusive because their 95% CIs cross the line of no effect. However, the white diamond, which summarizes all studies together, lies clearly to the right of this line and a dotted line is

projected vertically through the data for reference. By pooling the data, the effective sample size increases, resulting in a narrower combined 95% CI compared with those of the individual studies.

5. The pooled odds ratio from the meta-analysis is 1.51, with a 95% confidence interval (CI) of 1.32–1.74. Since this interval lies entirely to the right of the line of no effect, the result is statistically significant. The combined analysis therefore indicates that, across all studies, there is a 51% increase in the odds of success compared to the null hypothesis.

Statistical power

We noted that three of the five studies in the forest plot have a 95% CI range which crosses the line of no effect, but when all studies were combined, the meta-analysis produced a statistically significant result with a much narrower range of data. Much like adjusting the focus of a microscope, enlarging the sample population sharpens the picture, making it easier to detect small but meaningful effects within the data.

How does this work?

To understand this, we need to introduce the concept of **statistical power**. Remember from Chapter 3 that all studies carry the risks that a hypothesis may be falsely accepted (type I error, α) or that the null hypothesis may be falsely rejected (type II error, β). The risk for each varies between 0 and 1 with 0 representing no risk and 1 representing certainty.

Power is about the risk of type II error and is mainly influenced by the size of the effect you are measuring and the number of independent observations made in the study. The greater the statistical power the lower the risk of type II error, and so power $= 1 - \beta$. You can see this effect in the meta-analysis where small studies (small symbols) tend to fall on either side of the line of no effect with large CIs, whereas larger studies fall away from this line on the same side of the graph. Statistical power is calculated in different ways depending upon study design but, in principle, it is linked to sample size (N), the value of α (0.05), the effect size (r) and the probability of making a type II error (β) so if any 3 of these parameters are known, the other can be determined.

Heterogeneity

Funnel plots are often used to describe the precision of studies used in a meta-analysis (Figure B). By plotting the standard error as a measure of study size against the odds ratio (or the main measurement outcome from the meta-analysis), we can project a triangular funnel-shaped distribution of variance which comes to a peak at the top and spreads out to a broad base at the bottom. High-powered studies with low standard error cluster around the mean result of the meta-analysis towards the peak of the plot (A) with lower-powered studies tending to scatter more widely across the base (B). Since 95% confidence limits vary inversely with study power, the projected funnel-shaped confidence interval boundary helps to identify studies which might fall outside the natural variance produced by the meta-analysis (C). In Figure A, for example, the SHINE-I study has a large weighting due its sample

Figure B Funnel plots are often used to assess precision of studies (black circles) in a meta-analysis. The odds ratio is expressed as a log value to allow for linear comparison between studies.

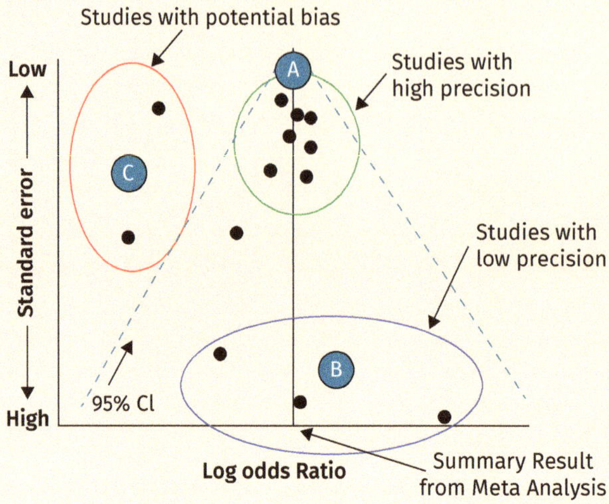

size with narrow 95% confidence intervals suggesting low variation around the mean. This is a high-precision study which would appear towards the peak of the triangle in a funnel plot. The CARE study produces a similar odds ratio but has a very small weighting due to a small sample size and large 95% confidence intervals. This is a low-powered (low-precision) study which would fall towards the base of the funnel plot triangle. Most clinical trials are conducted with high integrity but let's now imagine that one or two studies contained data where the experimenters had intentionally selected results that favoured their hypothesis and discarded negative results. This is called the **file-drawer problem**, where inconvenient data are excluded for no good reason from the analysis as if they had been hidden away in an imaginary file drawer. These actions would exaggerate the effect size and reduce the standard error of the study causing it to fall towards the top of the funnel plot but outside the confidence interval boundary (C in Figure B).

Overall, funnel plot analysis helps us to see how differences in precision between the studies in a meta-analysis affect the overall result. It also helps to highlight problem studies affected by bias or file drawer effects.

Pause for thought

- Clinical scientists conducting a meta-analysis typically begin by evaluating the quality and reliability of each study they plan to include. This initial assessment ensures that only robust and trustworthy evidence contributes to the overall analysis. Use your understanding of objectivity (Chapter 1), the statistical road map (Chapter 3), and Bradford-Hill's Causality criteria (this chapter) to think about criteria you would use to assess if a study is eligible for inclusion in a meta-analysis.

Cochrane Collaboration and the future of evidence-based medicine

Evidence-based medicine (introduced in Chapter 1) is the process of making decisions about healthcare treatment using unbiased, reliable research. This is difficult to do because humans are naturally biased, and conflicts of interest tend to get in the way of clear-headed decision making. What if you discovered a disease cure but stood to make a profit from promoting it to healthcare providers? Might you be tempted to put your cure ahead of alternatives which might work just as well or better?

The Cochrane Collaboration is a network of international organizations who share high quality, reliable data from clinical trials and meta-analyses without seeking to make profit. They make this data available to the public through the Cochrane Library and support independent assessment of study quality by providing free access to risk-of-bias computer tools and educational programmes. Now in its 30th year, the great data-sharing initiative of the Cochrane Collaboration has increased the rate of progress in developing treatments for many different diseases.

In this chapter we have tracked epidemiology from its birth to the present day. The lessons from Snow, Hill, Doll, and Peto point consistently to a single conclusion: larger studies yield more reliable evidence. In his 2015 State of the Union Address, US President Barack Obama announced a new era of scientific exploration in Precision Medicine. This is a new branch of epidemiology which uses genomic, transcriptomic, proteomic, metabolomic, and bioinformatic 'big data' approaches to provide medical care tuned to the individual characteristics of patients. It's tempting to speculate that this is the logical consequence of not simply results, but also methodological breakthroughs, made by the 20th-century pioneers of epidemiology.

Figure 5.14 The Cochrane Collaboration has provided open access to clinical trials data and systematic review tools for over 30 years.

Chapter summary

- Bias is a natural part of being human.
- Scientists throughout history have been aware of the problem of bias and have developed different methods to control for it.
- The science of epidemiology uses objective, unbiased methods to understand the causes of disease.
- Experiments can be designed to minimize bias using approaches like the Latin square, randomization, blinding, and independent data analysts.
- Hill's criteria for disease causation provide epidemiologists with a framework for understanding the origins and causes of disease.
- Large clinical studies like the British Doctors Study help to reliably inform the public about the causes of disease in a world where misinformation can be spread through biased advertising and social media.
- By adding the results of worldwide clinical studies together, we can use techniques like meta-analysis to increase statistical resolving power and to find meaningful small and large effects in the data.
- Public health statistics and epidemiology studies made available through organizations like the Cochrane Collaboration help to build a future where decisions are made with reliable data to cure disease and extend human lifespan.

Discussion questions

1. The Infected Blood Enquiry published online in the UK in 2024 revealed how communication between governments, healthcare providers, and scientists and the public can go very wrong. How might reliable statistics help to avoid misunderstanding and miscommunication between each of these groups?
2. In Chapter 1 we learned from Dongen and Sikorski's paper 'Objectivity for the research worker' that the researcher themselves is the major source of bias in a study. From what you've learned in this chapter, can you build a list of different ways in which bias might contribute to all stages of investigation, from setting a hypothesis to writing up a research study for publication?
3. Epidemiologists are sometimes asked 'what is the definition of good health?' Use your understanding of bias and epidemiology from this chapter to try to answer this question. You may find it helpful by first attempting to define 'what is disease?'
4. How might you design a randomized controlled clinical trial to test the hypothesis that vaping carries the same risk for heart disease as tobacco smoke? You might want to explore the Cochrane Library online to find if there are clinical studies exploring this question.

Further reading

Lind, J. *A Treatise of the Scurvy in three parts*, Wellcome Collection. You can see Lind's Treatise online.

James Lind's original discussion on the causes and possible cures for scurvy. You can read about his fair trial on HMS Salisbury on page 149 of the treatise.

Held L. Psychoanalysis shapes consumer culture, or how Sigmund Freud, his nephew and a box of cigars forever changed American marketing. *Monitor Psychology*, 40 (11): 32, 2009.

A short history of Edward Bernays and his influence on American culture and marketing.

Darby S. A conversation with Sir Richard Doll. *Epidemiology* 14: 375–379, 2003.

As part of the ***Voices*** feature of the journal, ***Epidemiology***, Sir Richard Doll is interviewed about his collaboration with Bradford-Hill and the events which led to the British Doctors Study.

Richard Peto on why smoking kills but quitting saves lives. The Life Scientific Podcast with Jim Al-Khalili. 2019 April 9.

Richard Peto describes his involvement with the British Doctors Study and the development of systematic review and meta-analysis.

Cochrane Collaboration: Cochrane at 30: what our 30th year can tell us about the future. Search online for the article on Cochrane's official site.

A collection of reviews which summarize some key achievements through the Cochrane Collaboration in the field of healthcare and biomedical science.

6 SHAPES, SCALE, AND YOU

Picture the people around you—their bodies, their proportions, their outlines. Then turn to the rest of the living world, where every kind of organism has its own distinctive form. You only need a silhouette to recognize whether an organism is a bird, an insect, a snake or mammal, a cactus or a deciduous tree. The fossils of some organisms, such as marine molluscs, dragonflies, and sharks, show us that their shape has not changed much over millions of years (Figure 6.1). The exploration of natural shapes has led to some of the most profound ideas in biology, from the theory of evolution to the search for the molecular clocks that create complex patterns in the body. This chapter looks at how the search for number patterns in nature has influenced our understanding of shape and the statistical methods we use to understand it.

Harmonic proportions and fractals

The story of shape and how to measure it takes us on a journey of how biology works in time and space. Find a tape measure and try this simple experiment:

1. Divide the distance from the top of your head to the ground by the distance from your navel to the same spot on the ground. Jot down the ratio on a piece of paper.

Figure 6.1 Shapes that last. These fossil shells from marine molluscs (A) are around 40 million years old, yet they look very similar to ones you might find on the beach today, while fossil dragonflies are often found that look just like this living example (B). Credit: Anthony Short

A

B

2. Now compare this to the length of your longest finger to your elbow divided by the length of your longest finger to the middle of your chest.
3. Finally, work out the distance from your left ear to your nose divided by the distance from your left ear to your left pupil as you look forward.

Now compare your ratios; you should find they are remarkably similar. You have measured a repeated **harmonic proportion**, which defines the shape and symmetry of your body. Just like a tree, your central body trunk is branched, creating arms and legs, which branch again to create hands and feet, and again to create fingers and toes. Even though your finger bones are much shorter than your arm bones, the molecular process of branching is tightly controlled so that bone size decreases at regular points in time and space.

The 20th-century mathematician, Benoit Mandelbrot, used the term **fractal geometry** to describe how natural shapes remain self-similar as size decreases in harmonic proportion to the overall shape. His famous image of the branching routine of the airway tubes of the lung (Figure 6.2) attempted to explain how the developmental cues or 'routines' shown in Table 6.1 control the number, size and shape of tubes at each branch generation:

Figure 6.2 Mandelbrot imagined a simple branching model of the lung where airway tubes become narrower at each branch generation (A). Linking the mid-point of tubes at each branch generation produces a fractal spiral like the one seen in (B) and the fossil shells in Figure 6.1A.

A

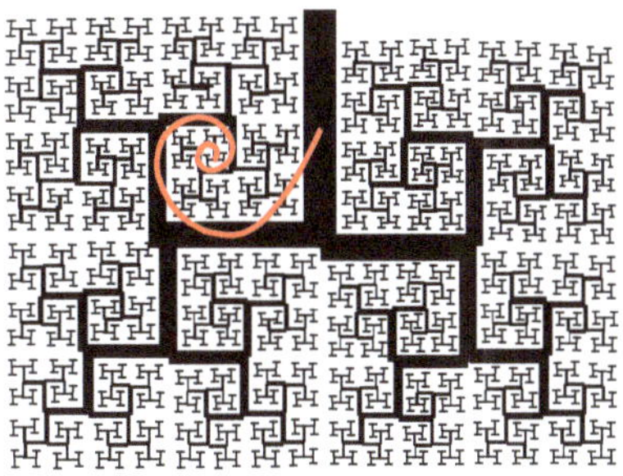

B

Table 6.1 The branching growth routine of the lung.

Branching Event	Geometric Event
Initiation	Establishes the point from which tube outgrowth will occur
Elongation	Determines the length and circumference of the tube at that branch generation
Stop	Determines the length of tube at that branch generation
Bifurcation	An event which splits the growth cue in space and reorientates it in preparation for the next growth cycle
Dampening of growth cues	Reduces the potency of the growth cue according to a harmonic proportion
Repeat	Cycle repeats at next generation

If you travelled down the path of an airway tube from the trachea to the alveolus, it would seem as if you were descending a narrowing spiral staircase.

This idea inspired the anatomist, Ewald Weibel, to measure the fractal branching routine of airway and blood vessels tubes as they synchronously branch in the human lung. He found that the structure of the lungs is the result of a replicated pattern of 23 branching cycles, with each cycle decreasing in size in harmonic proportion to the previous one. The result is over 8 kilometres of airway tubing packaged into a chest cavity with a 6-litre volume!

These fractal routines operate as geometric shape controllers, or **developmental clocks**, which define the harmonic proportions of all branched structures in your body, from your limbs, lungs, brain, to your blood vessels, kidneys, and reproductive system. Identifying the exact molecular signals which control this process remains one of the great challenges in developmental biology. You can find out more about this in another title in this series, *Animal Developmental Biology: Embryos, evolution, and ageing* by Julia Paxson.

The biology of proportion, size, and living harmonic frequencies has been studied for over 2000 years. Ancient philosophers were fascinated with the properties of shapes and studied them to find a single mathematical expression of geometric perfection and beauty. You can follow their line of reasoning using the process in Figure 6.4:

1. Draw a straight line and label one end A and the other end B.
2. Now think where you would cut the line (C) to create two segments that are unequal in length, but which are proportionately the same such that AC/AB=AC/CB, just as in panel (a). The position at which

this relationship occurs gives a value of 1.618 and is often referred to as the **golden ratio**, represented by the Greek letter phi (ϕ).

3 Use the long and short segments to create a rectangle as in panel (b). What do you notice? The ratio between the long and short sides of the rectangle equals ϕ as does the ratio of the long side to the furthest separated corners measured across the shape.

4 Now draw a line across the rectangle to join the two points at which ϕ occurs along the long and then short sides.

5 Repeat this process to create a pattern of rectangles with ever diminishing proportions but which all have fractal dimensions of ϕ.

You can take this further by drawing a curve between each segment of the rectangle to create a golden spiral which shows how proportionality can be translated from straight sided to curved structures as in panel (c).

A look at body structures like the shell of a nautilus or the repeated branching pattern of the lung shown in Mandelbrot's lung branching model (Figure 6.3) seem to capture the essence of this process and show us that the world of biology is abundant with variation in growth and form.

Figure 6.3 Plastic cast of the human airway (white) in the right lung and together with the pulmonary artery (blue) and vein (red) in the left lung. The branching follows the same harmonic proportions to form both the respiratory tree and the capillary network that conduct air and blood to the alveoli.

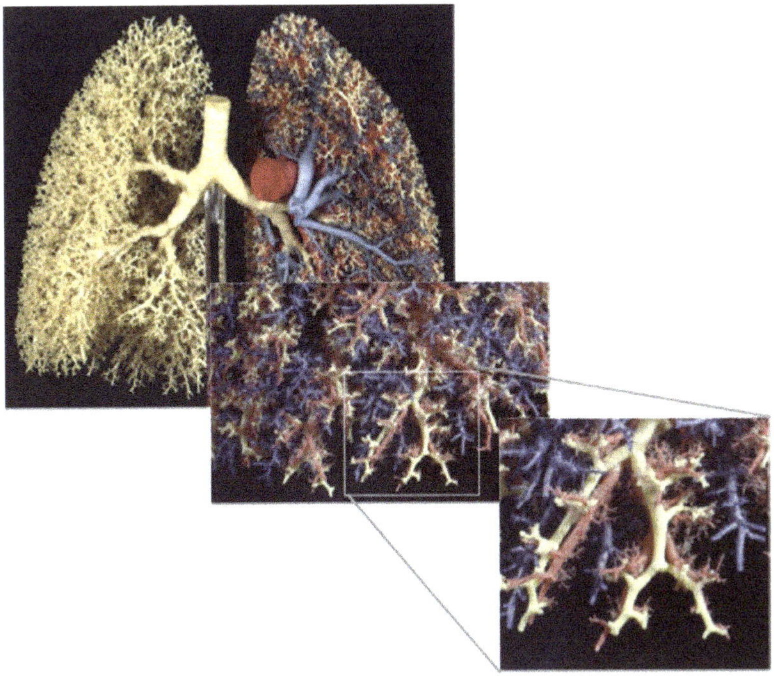

Figure 6.4 Experimenting with the golden ratio. The nautilus shell below (B) shows how regulated change in harmonic proportions can lead to beautiful biological structures.

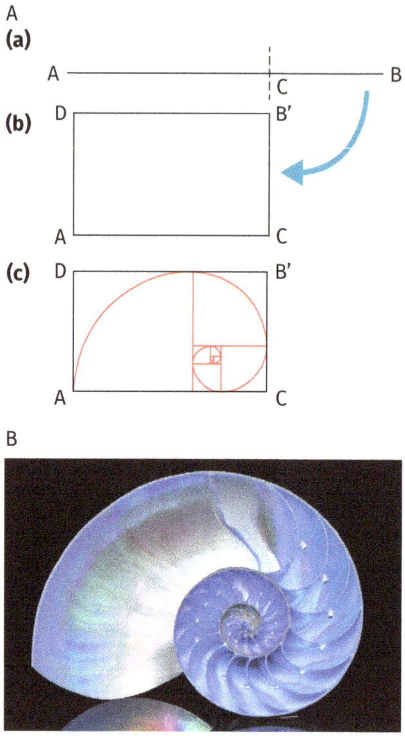

The golden ratio and the human body

Now look at the body dimensions you measured at the start of this chapter. Depending how you made the measurements, the proportions you measured may be roughly similar to 1.618, the golden ratio or ϕ. Vitruvius, a Roman architect, believed that the perfect human form should fit into a circle and a square with proportions between each major body segment that match ϕ. Leonardo da Vinci drew these proportions over 2000 years later in his famous image of Vitruvian Man (Figure 6.5). This link between geometric form and nature is captured in famous sculptures such as Michaelangelo's David and the Venus De Milo, and in famous buildings such as the Parthenon in Athens and the Eiffel Tower in Paris.

In this chapter, we are going to look at the experimental uses of harmonic proportions and fractals to explore the relationship between shape and function in biological systems.

From bee reproduction to the way you walk

Biologists have discovered that harmonic proportions do not just affect shape—it seems they also define the flow of genes and the population structure of communities.

The structure of a bee colony

A good example of how this works is found in honeybees (Figure 6.6).

A single queen bee produces all eggs in a colony. Sexually fertilized eggs produce female workers and new queens, whereas unfertilized eggs produce male drones. Males therefore have female parents, male and female grandparents, and female great-grandparents. Females, on the other hand

Figure 6.5 Ancient civilizations used the golden ratio to represent perfection in body proportions, for example in Leonardo da Vinci's image of Vitruvian Man. The golden ratio rectangle and spiral are superimposed over Vitruvian man to illustrate proportionality between body segments.

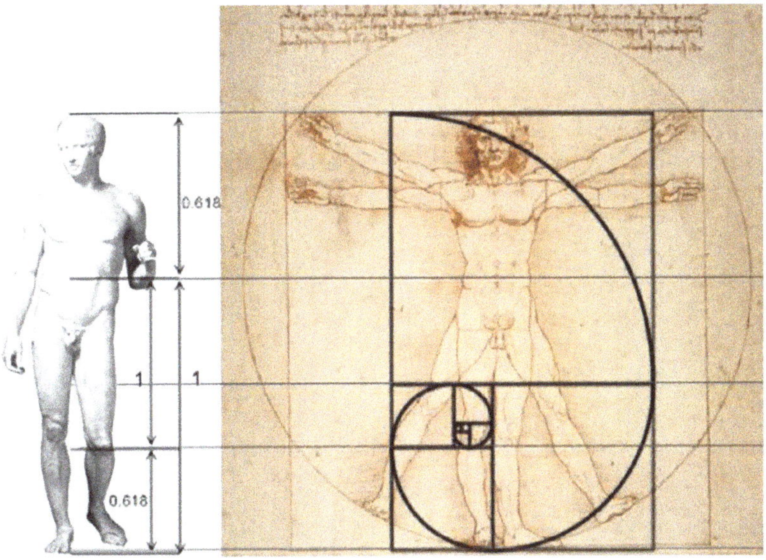

Figure 6.6 Female worker bees do most of the work in a colony.

always have male and female ancestors. If we follow the number of bees in the colony that come from a single male ancestor a harmonic number pattern appears (Figure 6.7). The bee colony increases as if the numbers in the two preceding generations are added to create the next generation, so:

$$0 + 1 = 1$$
$$1 + 1 = 2$$
$$2 + 1 = 3$$
$$3 + 2 = 5$$
$$5 + 3 = 8$$
$$8 + 5 = 13$$
$$13 + 8 = 21$$

. . ..and so on

This number sequence was first described by a 12th-century mathematician called Leonardo di Pisa, nicknamed 'Fibonacci', who used it to describe breeding in rabbits in his mathematical book 'Liber abaci'. Although his assumptions about the flow of hereditary traits in rabbits were wrong, they do apply to honeybee populations. The **Fibonacci sequence** follows the golden ratio because the product of each adjoining pair scales towards 1.618 or ϕ, as the numbers in the sequence become larger, so:

$$0/1 = 0$$
$$1/1 = 1$$
$$2/1 = 2$$
$$3/2 = 1.5$$
$$5/3 = 1.666$$
$$8/5 = 1.6$$
$$13/8 = 1.625$$
$$21/13 = 1.615$$

Figure 6.7 Expansion of honeybee populations follow a harmonic proportion like this one.

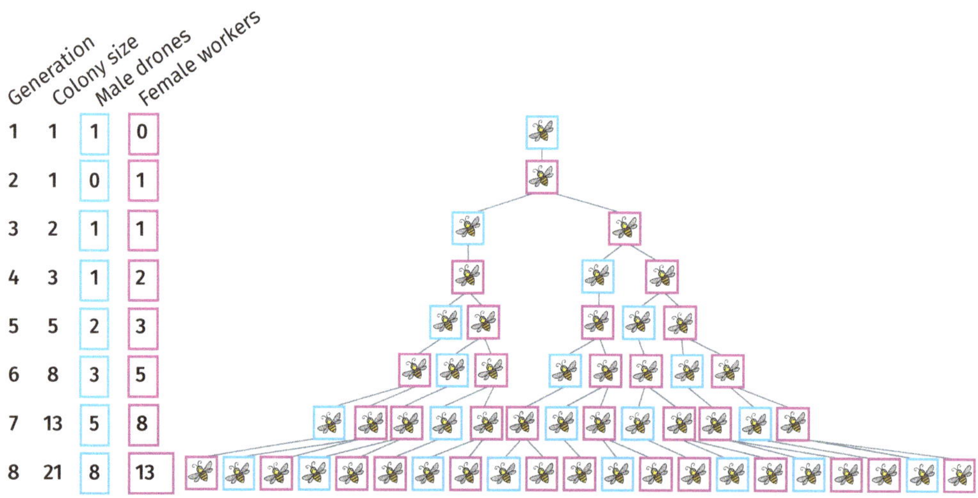

But why is this sequence important in biology? The golden ratio is an example of a harmonic proportion which describes how related things are controlled relative to one another, even if they might have quite different biological functions. In the beehive, for example, we know that the number of males to females follows the Fibonacci harmonic proportion and, by extension so will ratios of drone and worker tasks, genetic traits, nutrient demands, disease risk, and so on. There is no explanation of the significance of ϕ in beehive biology, but it seems reasonable to assume that it represents some optimal condition for colony stability.

This also seems to be the case for other biological systems because Fibonacci numbers appear in different contexts across the natural world, such as flower petals (3, 5, 8, and 13 are common; Figure 6.8), the number of clockwise and anticlockwise swirls of seeds on a sunflower head, the ratio of branch diameter changes in the veins of leaves, and the geometric swirl in a nautilus shell. Why not go to your local park and count how many natural structures you can find which have some multiple of a Fibonacci sequence within them?

Human walking

A second example can be found in the way you walk. Human walking involves two phases of movement which are linked by a harmonic proportion (Figure 6.9). The two phases are:

1. Stance (AC): begins when the heel first contacts the ground after a pace. It ends when toes leave the ground to begin the next pace.
2. Swing (CB): Begins when the toes have fully left the ground and the foot moves through the air. It ends as the foot strikes the ground after having moved through the air at the end of a pace.

Now that we have described two phases of movement, we can compare them to see if a harmonic proportion links gait to walking efficiency when we move from walking to trotting and running. For each phase of movement, the hypothesis is:

Stance/swing $= \phi$

Try measuring your own stance and swing phase movements and plot these on a graph against walking, trotting, or running speed. Does the

Figure 6.8 Fibonacci flowers—count the petals on the frogbit, apple blossom, and celandines!

A B C

Figure 6.9 A harmonic proportion links the phases of walking.

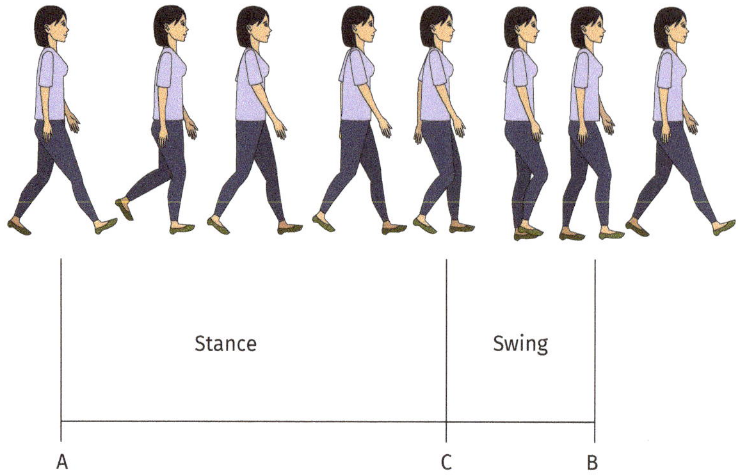

harmonic proportion of your gait match ϕ? Now close one eye and repeat the test (your depth perception will change so beware of trip hazards). Has the harmonic proportion of your gait changed as your mind becomes occupied with other information?

There are many neural processes that account for the rhythm of each gait—for example rhythmic neural connections between brain, spinal cord, and locomotor muscles. We know that commonly used drugs such as alcohol affect the brain—and affect the gait. Is it possible that if there are changes in the brain or nervous system due to illness, this could also affect the gait? Could the harmonic proportion of each gait be used to diagnose neurodegenerative disease?

This question has been investigated by researchers who have compared walking efficiency to the harmonic proportion in movement. They find that the most energy efficient harmonic proportion falls around ϕ and have called this the 'golden gait'. Patients with ataxia, a condition in which co-ordinated movements are disrupted, show a harmonic periodicity of movement which differs significantly from ϕ. Ataxia often accompanies neurodegenerative conditions such as Parkinson's Disease, and so detailed gait analysis may help pick up early symptoms of disease, when interventions and treatments are most effective.

Allometric relationships and the limits to body size and shape

Harmonic proportions of the type that Vitruvius and Fibonacci were interested in, and which Da Vinci later captured in his art, suggest that body shape and size matter a great deal in biology. It was Galileo Galilei (see Chapter 1) who first explored the relationship between geometric size, surface area, and volume. He noted that as objects become larger, the volume increases at a greater rate than their length and surface area. To understand this, we first need to look at how length, area, and volume are related to each other before thinking about the biology.

Take a cube with a side length, L (Figure 6.10). To calculate the surface area, imagine unfolding the cube to show all 6 sides laid flat. The surface area of each square is L^2 and there are 6 sides to the cube, so surface area of the cube, S, is $6L^2$. To calculate volume, imagine filling the cube with water. This fills the cube in 3 dimensions according to its height, width, and depth and so volume, V, is L^3.

If we increase the size of the cube, the side length increases in a regular straight line, or to the power 1, but its volume will increase to the power 3 so we can say that $L{\sim}V^{1/3}$. Another way to say this is $L{\sim}V^{0.33}$ because 1 divided by 3 is 0.33. Surface area rises to the power 2 and so the relationship is $S{\sim}V^{2/3}$ or $S{\sim}V^{0.67}$. For volume, $V{\sim}V^{3/3}$ or $V{\sim}V^1$. This relationship is called the **Square–Cube Law**, and can be used to predict length, surface area, and volume relationships in nature.

Allometry is the study of how the ecology, evolution, and physiology of all living things change in relation to body mass. It measures the change in a dependent variable over several orders of magnitude difference in body mass, perhaps from an amoeba weighing a few billionths of a gram, to an elephant of several hundred kilograms! To make sense of these big size differences, data are plotted on logarithmic X and Y axes and a straight line is regressed through the data points. The slope of each line is described by the equation, $Y=aX^b$, where b is the harmonic proportion, or **scaling factor**, that describes the relationship with changing size. Figure 6.11 shows how this works for length, surface area, and volume of a cube which scale to the power 0.33, 0.67, and 1 respectively. For a regular geometric shape like a cube these relationships hold true at any size magnitude and are described as **isometric**.

Figure 6.10 Length, surface area, and volume relationships of a cube.

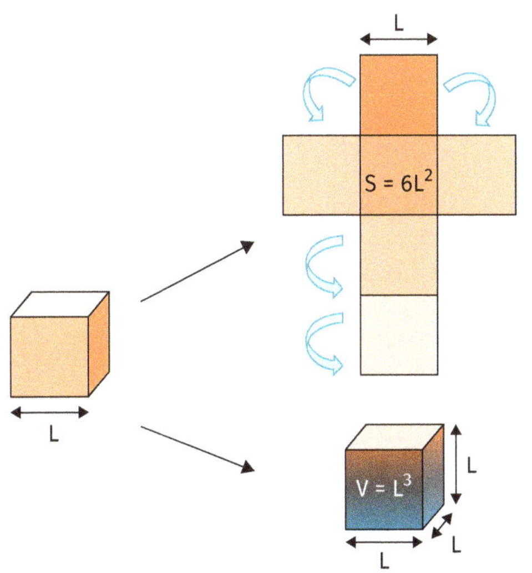

Figure 6.11 An allometric plot and line equations for volume, area, and length relationships of a cube.

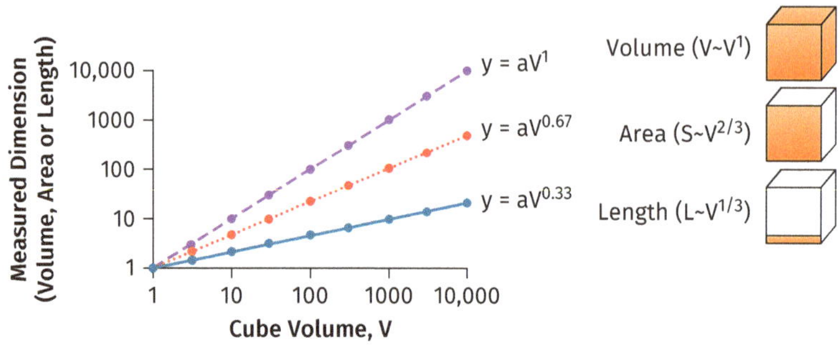

Galileo realized that scaling factors could help to explain how biological traits linked to length, surface area, and volume can limit the size of animals. Wonderfully, he asks the question, 'why are there no giants?' He reasoned that as size increases, so the length of a bone should be expected to change with body mass to the power 0.33 ($L \sim V^{1/3}$) but that tissues like muscle and tendon which attach to the bone surface should change to the power 0.67 ($S \sim V^{2/3}$). Following the square-cube rule, if a bone increased in length by 10 times so its surface area would have to rise by 100 and weight (remember that this is linked to volume) by 1000. There are no giants because, at a certain body mass, bones would become too heavy to be supported by muscle and tendon. Galileo explains this in his own words, excerpted from his book *Dialogues Concerning Two New Sciences* published in 1638 (Figure 6.12).

The square-cube rule has another surprising outcome because it suggests that power of bones and muscle becomes proportionately greater the

Figure 6.12 Galileo recognized that bone size and volume relationships place limits on body size as well as physiological performance.

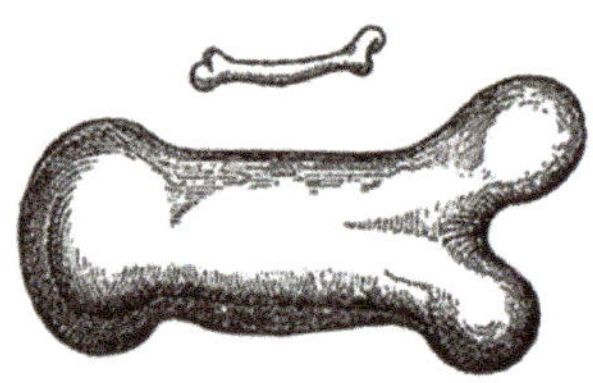

"I have sketched a bone whose natural length has been increased three times and whose thickness has been multiplied until....it would perform the same function which the small bone performs for its small animal. ... If one wishes to maintain in a great giant the same proportion of limb as that found in an ordinary man he must either find a harder and stronger material....or admit a diminution of strength......for if his height be increased inordinately he will fall and be crushed under his own weight. Whereas if the size of a body be diminished.....the smaller the body the greater its relative strength".

Galileo Galilei 1638
"Dialogues Concerning Two New Sciences"

smaller you become. It is for this reason that fleas, grasshoppers, and lizards can jump several times their body length, but large animals can only jump to a fraction of their own height (human, horse) or not at all (elephant). It explains why insects, such as leaf cutter ants, can carry loads that are up to 50 times their body weight whereas record loads for elephants of up to 500 kg amount to only one tenth of their body weight.

Case Study 6.1 uses allometry to look at how size affects top running speeds in mammals.

Case study 6.1

How do you solve the problem of being big and running fast?

Being big has lots of advantages. With a low surface area to volume ratio, it's easier to stay warm, store metabolic fuel, conserve energy, migrate efficiently over large distances, and defend against predators. Large size should also mean it's easier to run fast because stride distance and leg force against the ground are bigger than for small animals. But as size increases so gravity places more stress on the skeleton, so, to avoid the risk of breaking bones, big animals must either move more slowly, grow a heavier, stronger skeleton, or change the way their bodies move across the ground. Can we use allometry to understand how evolution finds a compromise between maximal running speed and body size?

A biologist, Dick Taylor, and an engineer, Christofer Clemente, asked whether differences in leg positioning in small and large animals might reveal advantages for running. They noted that small animals tend to crouch with legs folded beneath them, but large animals stand with limbs extended in a straight line (Figures A and B). If you're big, might there be a mechanical advantage to having long, straight legs?

Figures A and B: Small animals like this wood mouse on the left (A) have a crouching posture but large running animals like this horse on the right (B) stand on straight legs. Are these postures linked to size and running speed?

A

B

They used the ratio of the length of the leg extensor muscle arm (r) to the angle of force generated as the limb contacts the ground (R) to calculate a number called the Effective Mechanical Advantage (EMA). Think of this like a gear ratio that determines how easy it is to move across the ground. Have you ever watched a cat stalking its prey? The stalking phase of the hunt begins in a low crouched position with legs tucked under the body as the cat creeps slowly towards its prey. This posture gives a low EMA which allows for high contact and force generation against the ground with precise, slow hunting movements. As the cat springs forward to catch the prey it extends its legs and moves to long-legged running. This phase of the hunt has a high EMA for speed. Figure B shows how Dick and Clemente calculated EMA of different species of crouching and long-legged running mammals (Figure C).

Next, they used allometry to understand how EMA changes across a range of mammals with different body sizes and crouching or straight-legged postures. Figure D shows how EMA scales from chipmunks to horses over five orders of magnitude in body size. If we draw a line of best fit through the data, remembering that we are dealing with a logarithmic relationship, then EMA scales with body mass (M) to the power of 0.26 ($y=M^{0.26}$). In other words, as animals become larger so legs become straighter—but is this change in EMA enough to allow animals to run fast without risking damage to leg muscles, tendons, and bones?

The optimal relationship between length and volume is given by the isometric scaling factor $y=M^{0.33}$ and is shown in Figure D as a red dotted line.

Figure C Does posture affect running speed? Effective Mechanical Advantage (EMA) can be measured in crouching and upright animals as the ratio of the length of the leg extensor muscle arm (r) to the angle of force generated as the limb contacts the ground (R).

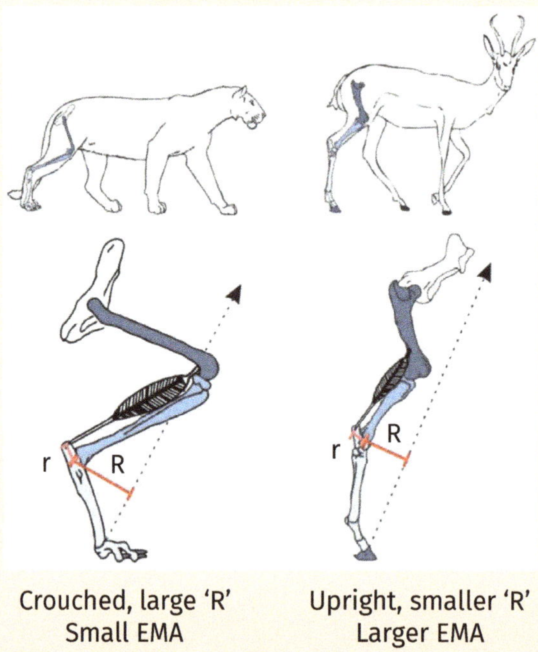

Figure D Effective Mechanical Advantage (EMA) scales over five orders of body mass magnitude with a scaling factor of $y = M^{0.26}$. The green shading shows how the mismatch with the isometric gradient for length becomes bigger with size.

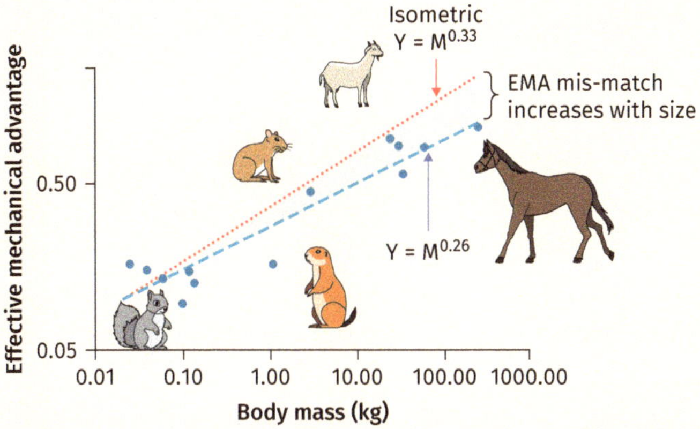

To keep the mechanical advantage of straight-legged running, the EMA–body size relationship would have to match this slope but, instead, we see that it is shallower and the difference between the two slopes, shown as a green-shaded gap, becomes bigger as body size increases. So, the mechanical advantage of running fast with straight legs is gradually lost as animals become larger, which leads us to ask, is there an optimal body size for running in animals?

In a classic allometric study, the evolutionary biologist, Ted Garland, tested this idea by comparing maximal running speeds of 107 mammals ranging in size from the tiny northern short-tailed shrew, weighing 16 grams, to the African elephant at 6000 kg. The results of his study are shown in Figure E. Garland found that maximal running speed does not produce a straight line with body size but instead, rises to a maximum, before beginning to fall. He found that an upside down 'u' shaped curve called a parabola fit the data much better and used the top of the curve to estimate optimal body size for running in mammals. This works out to be 119 kg, the size of an adult reindeer, with a maximal running speed of 80 kilometres per hour.

So, it seems that if you are large, a straight-legged upright posture is better than crouching for running fast, but at body masses above 119 kg the stresses on the skeleton become too great to support faster running speeds. This makes sense for animals like the reindeer or caribou which migrate long distances across the arctic tundra where food is scarce and the risk of predation from wolves is constant. Evolving the best possible body size for straight-legged running brings the benefits of being big, like staying warm and carrying metabolic fuel whilst lowering the risk of injury and inevitable predation. It also explains why very large mammals have evolved different strategies for locomotion. Elephants and rhinos have heavy skeletons to support a greater muscle mass but must move at lower speeds. Hippopotamus and the largest of all marine mammals, the whales, support their large body weights by living in aquatic environments. So as Galileo reasoned nearly 400 years ago, there are no giants, and if there were, they would be too weak to support their own weight.

Figure E Garland's allometric study looked at maximal running speed of mammals ranging in size from shrews to elephants. The dashed line is the allometric relationship which scales with $y=M^{0.17}$ but Garland found that the parabolic solid line fitted the data with greater accuracy. He used it to suggest that optimal size for running in mammals is 119 kg. The animals in the red box are ungulates like this white-tailed deer. He also showed that non-running mammals like sloths (S) are displaced far below the curve.

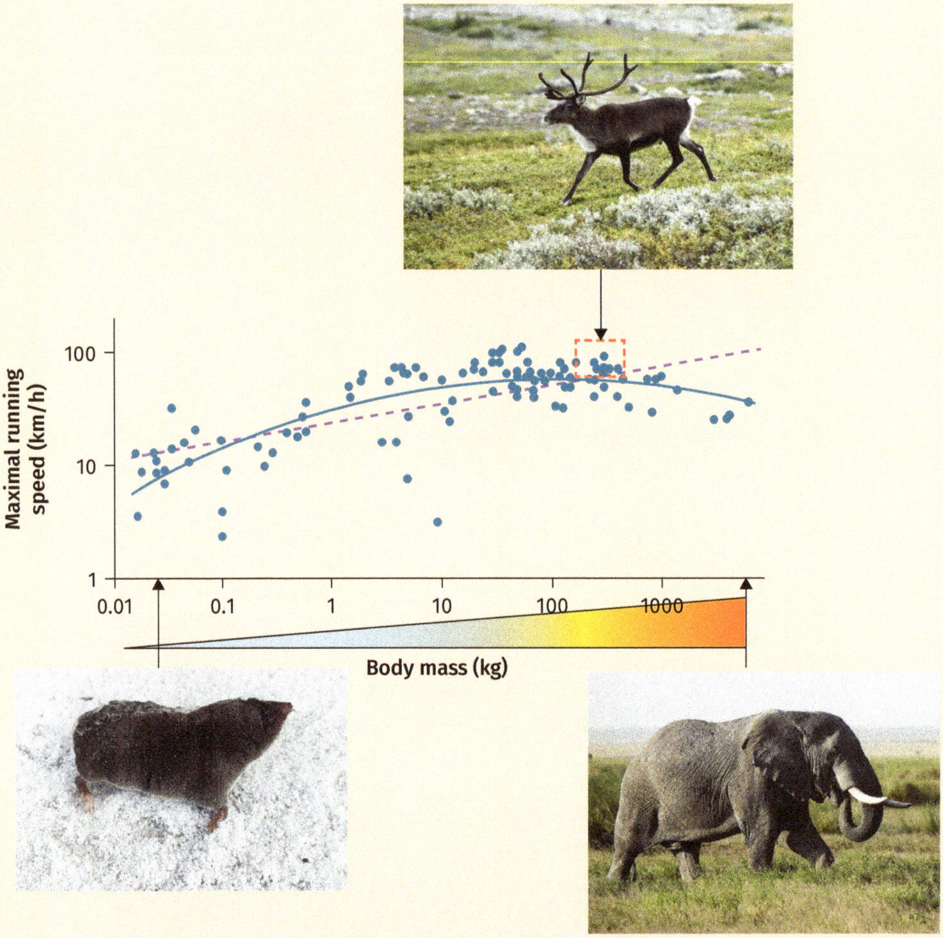

Pause for thought

Garland's discovery of an optimal body size for long distance running in animals implies that the process of allometry works at every level, from genetics to the biochemistry, anatomy, and physiology of growth and movement.

- Can you think of subcellular processes or structures whose abundance or function might scale with body size? Here is a question about mitochondria, the energy powerhouse of cells, to get you started: will one gram of mouse muscle contain the same, or more, mitochondria compared to one gram of elephant muscle?

Using statistics to look at the evolution of body structures

Body mass sets important limits for body structure and function, but it doesn't explain why body shape can vary so dramatically between closely related species. The Scottish biologist and mathematician, D'Arcy Wentworth Thompson, attempted to explain this by showing how body shape can change to suit different environments by altering or 'warping' certain dimensions. From Mandelbrot's ideas about fractals in nature (Figure 6.2), we know that this can occur because of small molecular changes in the developmental clocks which control body shape. So, for example, a basic body design can give rise to extreme shapes by changing the 'ticking speed of the clock', so that some structures grow more quickly whereas others may shrink or stay the same. Thompson described this using a map-like grid to illustrate how position of body structures changes between related species in different habitats. Figure 6.13 shows this both in fish and in several mammals—including ourselves.

This process is known as **morphological warping** and is used to explore how and why divergence evolves in physical structures. The idea of using a grid to show the morphing process presents a problem, however, because, just as with any countryside map, the larger the scale, the greater the detail we can see. How do you choose a scale that shows differences in shape, but which won't bury you in data?

Principal Component Analysis (PCA) is a statistical method which simplifies large data sets by reducing the number of variables (in this case, map co-ordinates) but preserves as much information about the relationships in the data as possible. It does this by standardizing variables to a common scale and then looking at how far they vary, or 'warp', from the mean and each other (multi-variance). Rather than reporting numbers to do with size or shape, this non-parametric approach ranks sources of variance between data points as **principal components** (see Figure 6.13).

Using allometry and PCA to understand the origins of canine tooth shape in sabre-toothed cats

Let's look how we use principal component analysis (PCA) to identify patterns that explain the evolution of body shape. Sabre-toothed cats were large predators which became extinct at the end of the last ice age, around 8-10,000 years ago. Their most obvious feature was a set of long canine teeth which extended well below the lower jaw (Figure 6.14). They look formidable but make little sense as an adaptation for hunting. Imagine how wide a mouth would need to open to clamp onto the neck of a running antelope. Then there's the muscle attachments—these would have to run at such a shallow angle to the jaw that the bite force generated would be barely strong enough to break the hide of the prey. These apparent contradictions have led palaeontologists to question whether sabre teeth were adaptations for hunting at all. Might they have been much more important for communicating social dominance and sexual fitness?

Figure 6.13 Warping of body structures produces different body shapes in nature. A) Dorsal-ventral stretching of the reef dwelling *Diodon* (porcupine fish) leads to a body shape like the *Mola* (sunfish) that is suited to an open ocean environment. B) Warping produces the different skull shapes of the human, which favours brain size, and the chimpanzee, which favours jaw strength. Adapted from D. Wentworth Thompson, On Growth and Form.

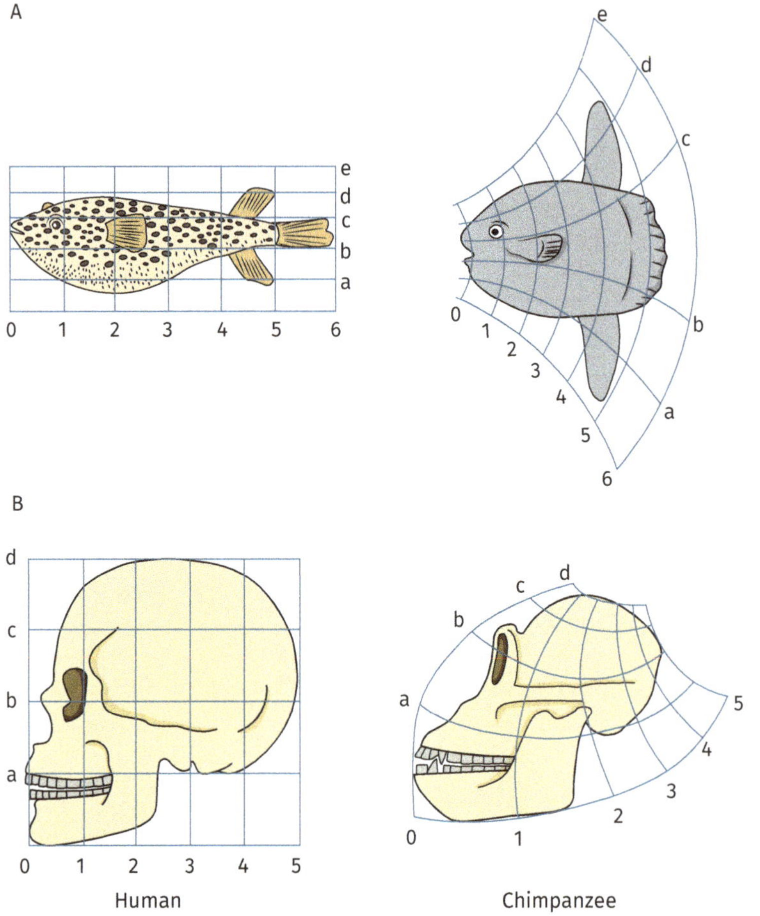

Figure 6.14 (A) Fossils from famous sites like the La Brea Tar Pits in California allow palaeontologists to use art reconstruction to gain an idea of how sabre-toothed cats must have appeared at the end of the last ice age. (B) This sabre-toothed cat skull shows the formidable hyper-extended canines.

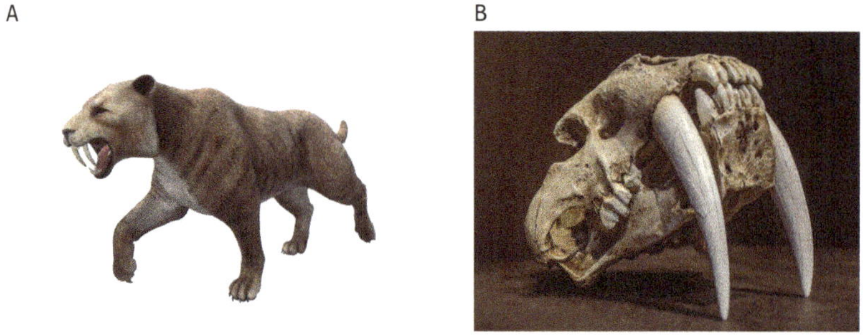

The sabre-toothed trait has evolved at least four times in different mammal species but tends to converge upon two types of tooth:

- a long dirk-like fang up to 21cm long which was smooth and flattened like a knife (a dirk is a long dagger used historically by Scottish warriors in battle)
- a shorter scimitar-toothed form which was rough and serrated.

Figure 6.15 shows these in comparison to the conical tooth shape we now see in living cat species.

If dirk and scimitar canines evolved only as adaptations for hunting through **natural selection** then we would expect that the variation in tooth shape across species would be like the carnassial teeth which are those at the back of the jaw and only play a role in cutting and chewing. If, however, they evolved as ornaments to communicate sexual fitness and social then their shape variation will be driven by **sexual selection** and should differ statistically to teeth that only play a role in eating.

We can use principal component analysis to compare variation in the shape of canine and carnassial teeth to test the following null hypothesis:

H_0: the variance in canine tooth shape for sabre-toothed cat species will not differ from that of carnassial tooth shape, an indirect measure of natural selection.

A simple two-dimensional x-y axis graph could be used to explore tooth shape relationships two at a time, but this only tells part of the story. What if combinations of other shape dimensions influence the distribution of the data much more than just the two compared on the graph? PCA allows us to interpret the effect of several dimensions on a trait whilst minimizing loss of information.

Figure 6.15 Comparison of dirk, scimitar, and modern-day conical-toothed cat skulls.

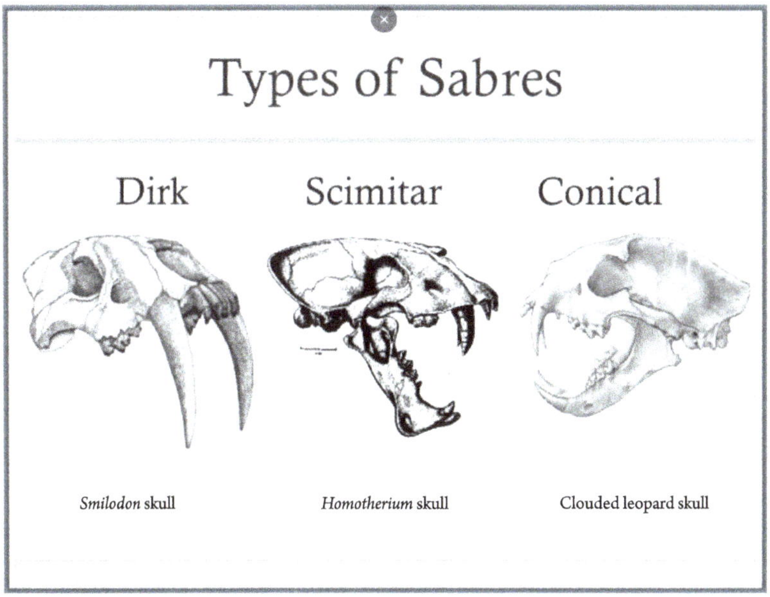

Figure 6.16 Tooth and skull dimensions used to explore the evolution of sabre-toothed and conical-toothed traits in extinct and living cat species. CW: canine width; CL: upper canine length following gumline; CCH: upper canine height from gum to tooth tip; P4L: upper carnassial length along gumline; M1L: lower carnassial length along gumline; SL: skull length. Adapted from Randau et al., *Pone* 8(8): e72868, 2013.

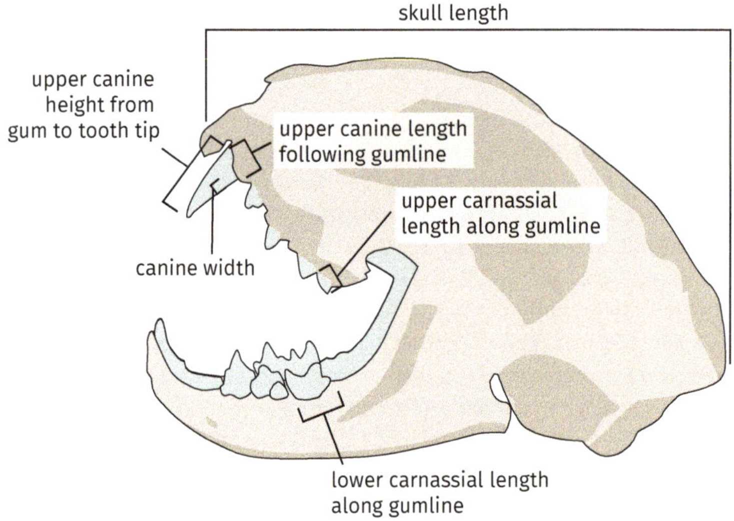

Figure 6.16 shows the tooth and skull length measurements used, with each measurement giving a shape dimension that will change with natural or sexual selection. Information about tooth width, height, and length for extinct sabre-toothed and living cone-toothed cats can be drawn from fossil tooth and skull bone records in museums around the world. We cannot calculate body mass for extinct species due to incomplete skeleton records and so a ratio of tooth size to skull length is used to correct for differences in body size. The measurements to be made are:

- Upper carnassial P4L: Skull Length (SL)
- Lower carnassial M1L: SL
- Canine Crown Height (CCH): SL
- Canine Length along the gumline (CL): SL

To begin the PCA, each tooth shape dimension is determined as a proportion of skull length (SL) and then converted to a scaled index. We do this by subtracting each data point from its mean tooth dimension and dividing by the standard deviation. This removes any effect that differences in scale might have on the data but does not alter the relationship (distance) between each data point. This is a key step in PCA because it allows comparisons to be made between factors that might vary greatly in size.

Figure 6.17 shows a biplot of the scatter, or data variation, for each of the tooth measurements as ratios to skull length for sabre-toothed and living conical-toothed cat species. Rather like a set of map co-ordinates and a compass angle, **eigenvectors** describe the direction of variation with respect to each measurement, whereas **eigenvalues** tell us how the variance falls in one direction compared to another. In this way, eigenvectors and eigenvalues behave as scaling factors which determine the position of each data point based on the influence of each tooth shape measurement. Differences in variation between the measures falls to zero where the eigenvector lines converge and so data points falling here align to the null hypothesis: that tooth shape variation is the same for all measurements.

So does the biplot data in Figure 6.17 suggest sabre teeth are about hunting or sexual fitness? Most of the data scatter along the eigenvector for canine crown height: skull length, CCH:SL, and so this relationship is called principal component 1 (PC1) because it contains most of the variation for tooth shape and skull size. The next greatest variation is found in the ratio of canine height to skull length, CH:SL, giving principal component 2 (PC2) whereas the smallest variation in the data is associated with the carnassial tooth measurements, M1L and P4L, giving principal components 3 and 4. The angles of the eigenvectors tell us about the difference in the direction of variation in the data. From this, we can tell how closely correlated the measures of tooth shape are to one another.

- Where eigenvectors follow the same angle, as for the two carnassial principal components PC3 and PC4, there is little difference in the variation and, so, these measures are closely correlated.
- Where the angles are greater, the direction of variation for each measure becomes increasingly independent from one another. This becomes interesting if we compare our measures of canine shape, PC1 (CCH:SL) and PC 2 (CL:SL), to measures of carnassial shape, PC3 (M1L:SL) and PC4 (P4L:SL).

The differences in eigenvector angle tell us that canine shape is weakly associated with carnassial shape, our measure of natural selection. As we move along PC1 away from the point of convergence between all eigenvectors we encounter data points for living conical-toothed cat species to the

Figure 6.17 Plotting the ratio of upper canine crown height to skull length (CCH:SL) against P4 upper carnassial length to skull length (P4L:SL) produces a scatter of data points across different data ranges for each measure made in a different species of cat. Scaling places each measure onto the same scale without affecting the scatter (variance) between the data points.

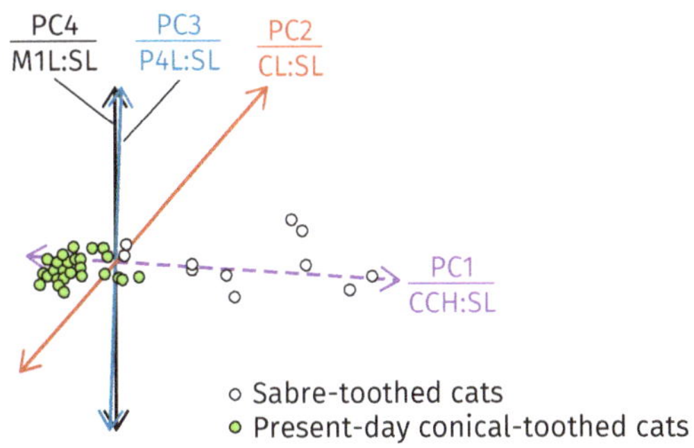

left and sabre-toothed cat species to the right. We can conclude that variation in canine shape is highly divergent between conical- and sabre-toothed species and that this effect is not driven by hunting success, a process of natural selection. In the absence of other measures of selection, it seems that the sabre-toothed trait evolved as a symbol of sexual fitness and group dominance.

Since most of the variation in the tooth shape data is contained within PC1 and PC2, we can use this information to explore evolutionary relationships between living conical-toothed cat species and extinct sabre-toothed cats. Figure 6.18 shows a plot of the PC1 and PC2 tooth shape data where data points fall into three distinct clusters with each ellipse indicating the boundary of the 95% confidence limits for each cluster. Most sabre-toothed cats cluster separately from others but notice how some fall within a middle cluster that includes living conical-toothed species. Might this cluster be a living link to the sabre-toothed cats of the past?

A **dendrogram** is a branching diagram that shows the closeness of eigenvalue points within a PCA. Each eigenvalue represents a different cat species and so we can use this to think about evolutionary relationships between our cat species based on the two principal components used in the cluster diagram. Figure 6.19 shows a dendrogram with the species names for sabre-toothed cats shaded in pink and living conical-toothed cats shaded in green. It shows us how two sabre-toothed species seem to cluster differently among living cats and suggests the sabre-toothed trait has evolved in different ways and at different times.

PCA is a powerful statistical tool that can be used to explore many kinds of biological relationships and which opens the door to new lines of thinking about evolution and natural and sexual selection. So the next time you look at your pet cat, remember you just might have a sabre-toothed tiger in your living room!

Figure 6.18 Cluster plot showing three clusters of tooth shape when we compare living conical-toothed cats to extinct sabre-toothed species. Might the presence of conical- and sabre-toothed species in one cluster suggest that some living cats share tooth shape characteristics of extinct sabre-toothed species?

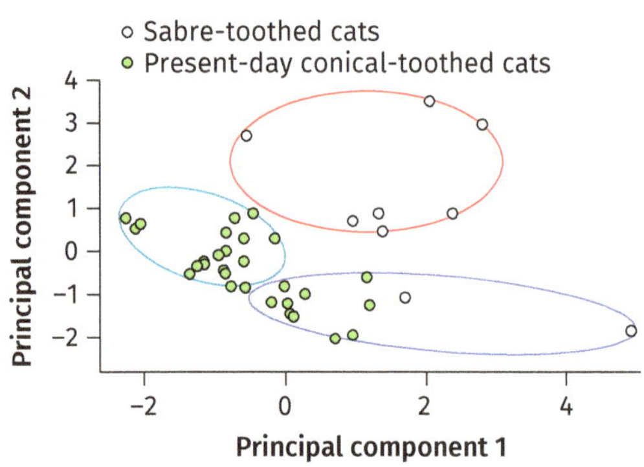

Figure 6.19 A dendrogram uses tooth shape to show how sabre-toothed (pink) and present-day conical-toothed cats (green) evolved from a common ancestor. Two sabre-toothed species seem to have evolved sabre-toothed characteristics in a different way to others.

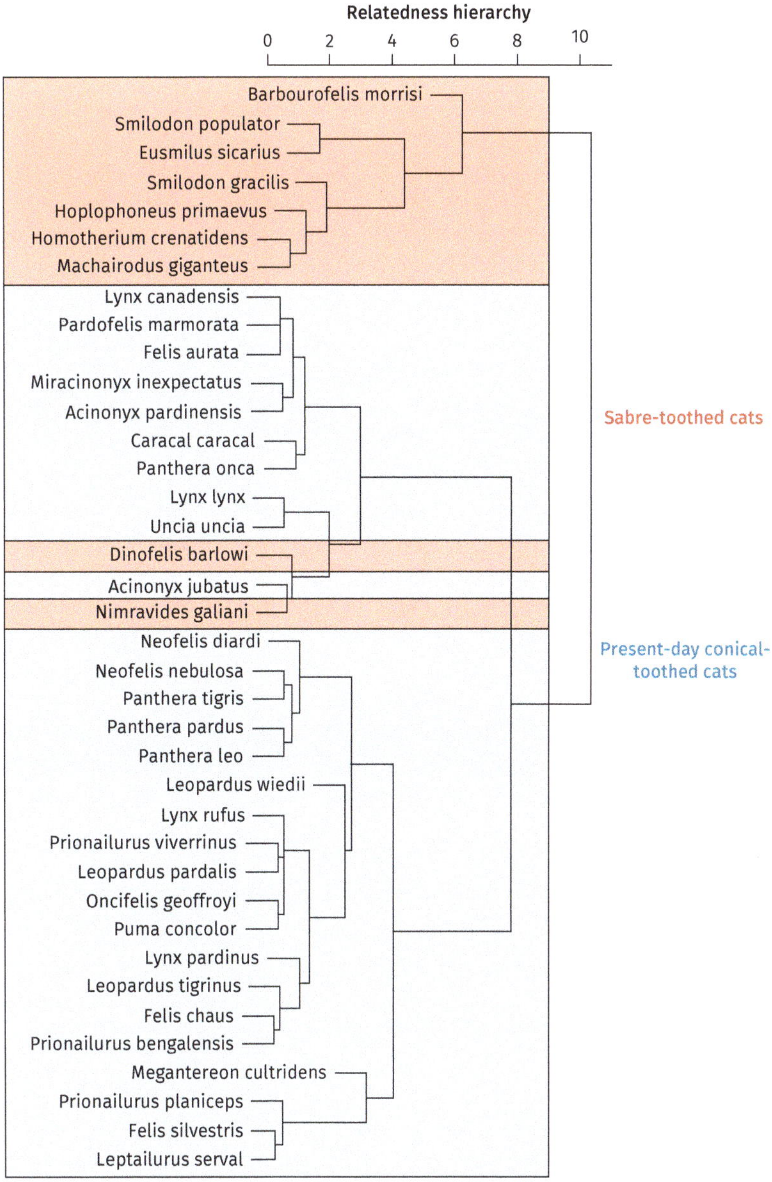

Modern-day sabre-toothed mammals

The sabre-toothed trait has evolved for different purposes in living mammals. Small forest-dwelling deer like the Asian muntjac and the African chevrotain (mouse deer) have hyper-extended canine tusks rather than antlers to establish social dominance and fight for mates. This adaptation makes sense for a life spent in thick forest vegetation. Some living mammals with long canines help us understand how sabre-toothed cats may have hunted with such large teeth. The South American short-tailed opossum is a small marsupial with hyper-extended canines which, in proportion to skull size, fall within the range of sabre-toothed cats. In this case, PCA has been used to show that the scimitar-like canines have co-evolved with changes in skull shape to support a wide gape and increased forelimb strength and dexterity. Here, the forelimbs are used to subdue and position the prey before the canines are used to deliver a killing bite to the neck then disembowel prey. Importantly, this adaptation allows the opossum to catch prey animals which are larger than itself. Taken together, the driving forces favouring the sabre-toothed trait can evolve through both natural and sexual selective processes in herbivorous as well as predatory mammals.

Shaping our future world with statistics and quantum computers

As an intelligent ape species, we have used the best understanding of the ages to explain the world around us for at least 5–10,000 years. A breathtaking illustration of how common scientific ideas can span millennia is kept in the British Museum in London. Figure 6.21 shows a 4000-year-old clay tablet from ancient Mesopotamia (now modern-day Iraq) which is moulded into the shape of a liver. You can clearly see the gall bladder and the hepatic portal veins which are surrounded by a regular grid and a series of holes. Look closer and you will notice small score marks inside each square that have been made using papyrus reeds. These are cuneiform marks. Cuneiform was developed as a common communication code to allow information to be transferred between peoples of different languages. It is thought that

Figures 6.20 Living sabre-toothed mammals! A) The African chevrotain (mouse deer) lives in thick vegetation and has exchanged antlers for sabre-tooth tusks to show sexual fitness. B) The South American short-tailed opossum has evolved hyper-extended canines which together with skull and forelimb adaptations are formidable hunting adaptations—in this case it looks like natural rather than sexual selection is the driver of evolution.

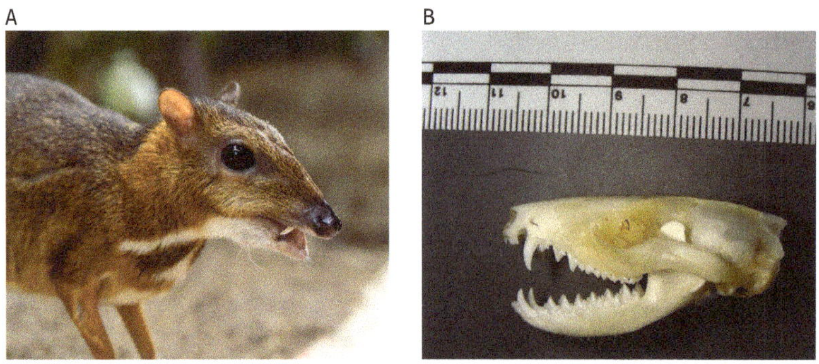

Figure 6.21 A 4000-year-old liver-shaped clay omen tablet from ancient Mesopotamia has the basic hallmarks of a modern computer.

omen tablets like this were used to interpret blemish marks on the liver of a sacrificed animal to make decisions about the future. Blotches in different areas of the liver were interpreted using the cuneiform information in each grid on the tablet. The purpose of the hole in each grid is uncertain but may have been to hold a peg so that blemish patterns and cuneiform marks could be mapped together across the liver. This method for predicting future events might seem ridiculous today but, looked at from a different perspective, the omen tablet has all the hallmarks of an early computer. It has input (liver blotch patterns), output (grid prediction matrix), an information storage system (holes for pegs), and a universal language code (cuneiform writing). Artefacts like this tell us that ancient peoples understood the principles of advanced technologies, but the scientific understanding of the age prevented them from developing their ideas in a meaningful way. In this last respect, we are no different from curious people who lived in past ages.

Each chapter in this primer ends with a discussion about the future of life sciences, from the search for objectivity to big data, omics, artificial intelligence, and meta-analysis. When we pull this together, the emerging picture looks exhilarating but hugely complicated! Figure 6.22 shows how true understanding of human life can be seen as a hierarchy of complex interactions, from genes to cells, tissues, organs, bodies, and communities. We learned from Chapter 3 that each of these structures is defined by its own genomic, transcriptomic, and proteomic interactions and, together, operate through biochemical reactions which work on a biological timescale of nano- to microseconds (10^{-9} to 10^{-6} s), through to a human lifetime of tens of giga seconds (10^{9}s). We know that each of these interactions will have its own shape or 'kinetic' character so we can expect that harmonic fractal, allometric, and clustering behaviour will be distributed through these hierarchies. At the level of communities, epidemiology seeks to understand how societies might be better organized to limit disease and promote human welfare. Together, these forces define what it is to be alive! Statistics and

Figure 6.22 Quantum computers apply statistical methods to reveal the connections within and between each level of life complexity from molecules to societies.

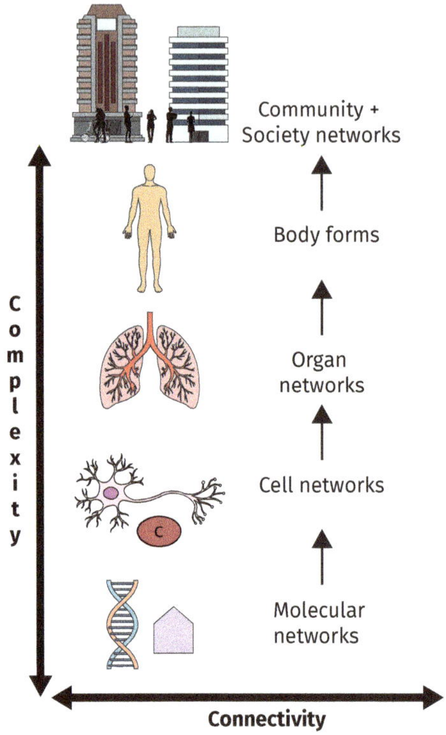

experimental design allow us to explore these interactions bit-by-bit, but how do we piece together the jigsaw puzzle to build a complete picture of life?

The experimental and statistical approaches of the future need to deal with scale, time, shape of interactions, and big data. Artificial intelligence and machine learning are revolutionizing our understanding of these interactions, but vast computing power is needed to process the statistical calculations that are needed to make sense of these processes. Enter the age of quantum computers. Unlike normal computers which process bits of information as values of either 0 or 1, quantum computers use subatomic quantum mechanics to process data as **qubits** which can exist in different states, giving many more bit values than just 0 or 1. A powerful type of statistical correlation called **entanglement** can be used to process multiple qubits at the same time. This speeds up computing time and allows for increasingly sophisticated statistical modelling on massive amounts of data. Combining principles of objectivity, experimental design, and statistics with AI, quantum computing, and modelling of complex systems is the next big leap for life sciences.

Chapter summary

- Harmonic proportions define the fractal patterns which give shape to body structures such as limbs, lungs, and other organ systems.
- Growth cues which create fractal patterns in our bodies work like molecular clocks to give regularly repeated shapes at different orders of size.
- Ordered number sequences, like the Fibonacci sequence, give shape to animal and plant forms but also appear elsewhere in life from the phases of movement to the structure of insect communities.
- Allometry uses a harmonic proportion called a scaling factor to describe how body structures change with size.
- When applied across several orders of size magnitude, allometric scaling factors help us understand the evolution of structure–size relationships in living things.
- Statistical methods used to analyse proportion and shape need to cater for the fact that increasingly finer scales give rise to greater detail in the data.
- Principal Component Analysis (PCA) simplifies large data sets by standardizing them to a common scale and measuring the distance between points.
- We can use this approach to look at the clustering of common traits which gives insight into their evolution and relatedness between species.
- Statistical approaches like PCA can be used by quantum computers to search for patterns in biological relationships at any order of size and time magnitude and represents the future of biostatistical analysis.

Discussion questions

1. Discuss the typical plants and animals you might see if walking through a park, forest, or along a rocky shoreline. How many of these will have structures that show fractal characteristics? Can you think of ways in which these harmonic proportions might help the animal or plant to survive and reproduce in its environment?
2. Some animals have evolved exaggerated body structures through the process of morphogenic warping. Think, for example, of the long neck of the giraffe or the extended pectoral fins of flying fish. Where might these sit compared to other animals on an allometric regression of neck length or pectoral fin length to body size? How might we use allometric relationships to explain the evolution of body shape to different environments?
3. Scientists are developing a new generation of quantum computers to tackle some of the most complicated questions about life. To understand the scale of this task, think of a simple biological observation and list the different levels of complexity that would have to be measured to understand this process from start to finish. For example, why do we have five fingers and toes, rather than a different number of digits?

Further reading

Ball, P. *Patterns in nature: Why the natural world looks the way it does.* University of Chicago Press.2017. ISBN: 9780521319874

A wider look at harmonic patterns and their forms in nature, from the shape of rivers to the structure of galaxies.

Nielsen, KS. *Scaling: Why is Animal Size so Important?* Cambridge University Press. 2008. ISBN: 9780521319874

A classic and very readable discussion of all aspects of allometry from the function of enzymes to the evolutionary constraints of size on animal life.

Thompson, DW. *On Growth and Form: the complete revised edition.* Dover Publications Inc. 1992. ISBN: 9780521319874

D'Arcy Thompson's classic book about the evolution of animal shape and form. Although written in 1917, many of the themes covered in the book are relevant to modern biology today and provide an important context for understanding how to explore body shape and size.

StatQuest: Principal Component Analysis (PCA), Step by Step with Josh Starmer. YouTube Video.

A very clear explanation of the idea behind principal component analysis set in the context of cell biology.

GLOSSARY

A

Abductive reasoning a method of generalized thinking where observations are used to draw the simplest possible explanation as to how something might work. In abductive reasoning, there is often room for doubt and sceptical challenge.

Allometry the study of how anatomical structures and physiological or biochemical processes change in relative proportion to body size.

Analysis of covariance (ANCOVA) a regression model which includes categorial and continuous variables.

ANOVA (Analysis of Variance) a statistical approach used to compare variance across the means of three or more sample groups.

Artificial intelligence technology that allows computers and machines to perform tasks in a way that is similar to the human thought process.

B

Bell-shaped curve a descriptive name for a normally distributed data set on a graph.

Binomial distribution table a distribution of probabilities which are used to assign significance for the Sign Test for one region of the interactome with another.

Bonferroni α Adjustment Method a type of post hoc test used in ANOVA which works by dividing the critical level of significance (α) by the number of groups to be compared.

British Doctors Study the world's first large-scale longitudinal study of a population of smokers which established the statistical evidence which links the effects of smoking to cardiovascular disease, lung cancer, and lowered life expectancy.

C

Causation issue of whether a change in one variable drives change in another.

Central tendency a single value which is used to describe the mid-point of a normally distributed data set.

Chi squared test a statistical test which is used to determine if observed data differs from expected data by chance.

Clustering the tendency of some parts of an interactome to interact more frequently with one another.

Cochrane Collaboration an independent, not-for-profit organization that produces and disseminates systematic reviews and meta-analyses of healthcare interventions.

Common-cause variation a natural and expected type of variation that is always present and inherent to the system which is being studied.

Comorbidity the presence of several health disorders in a person at the same time.

Confidence interval a measure of study precision given as a range of values that would be expected to contain the true result if the study was replicated several times over.

Control variable an effect which does not respond to manipulation of the independent variable by the experimenter.

Covariate a continuous independent variable which has a statistical relationship with a dependent variable and so can influence the main statistical comparison which is to be made. Covariates do not form part of the main experimental manipulation but may nevertheless influence the behaviour of the data.

D

Data distribution a graphical presentation of data from a sample or population.

Deductive reasoning a thinking process where logic is used to progress an idea from a general series of statements to a specific conclusion.

Deep learning a subset of machine learning which teaches computers how to replicate decision making in a way that is similar to neural networks in the human brain.

Degrees links between nodes in an interactome.

Degrees of freedom the number of independent observations in the data which are free to vary.

Dendrogram a tree diagram which is used to show taxonomic relationships.

Dependent variable an effect which is predicted by the hypothesis to respond to manipulation of the independent variable by the experimenter.

Developmental clock a set of molecular signals which operate during development to control the proportions of body structures at different orders of scale.

E

Eigenvalue a scaled dimension in a principal component analysis.

Eigenvector a linear proportion which describes the direction of variation amongst the principal components within a principal component analysis.

Eligibility matrix an assessment of bias risk among papers which are to be included in a systematic review or meta-analysis.

Empirical or 68–95–99.7 rule a statistical law in which standard deviations are used to explain the predictable distribution of data either side of the mean in a normally distributed data set.

Entanglement a form of correlation used in quantum computing that is being developed to advance the speed and complexity of statistical modelling and its use in artificial intelligence.

Epidemiology the study of causes, occurrence, and distribution of disease and how these affect the health of a population.

Eugenics a discredited method of study into the methods that might be used to select desired heritable traits in human populations.

Evidence-based medicine a systematic use of scientific evidence gained from multiple sources that is used to help healthcare professionals make decisions about the best way to treat patients and prevent disease.

F

Factorial design a statistical method which analyses the relationship between multiple independent variables simultaneously.

Fibonacci sequence a series of numbers in which each number is the sum of the two preceding numbers. The ratio between each Fibonacci number equals the golden ratio, 1.618.

File-drawer problem the selective reporting of scientific results (the name refers to hiding undesired results in an imaginary 'file drawer').

Fisher's least significant difference (LSD) a type of post hoc test used in ANOVA which looks for pairwise differences between groups in ANOVA using a variation of the t-test method. A variation of this test known as the Fisher-Hayter procedure adjusts for the risk of type I error.

Forest plot a graph displaying summary results and confidence intervals of multiple studies in a meta-analysis.

Fractal geometry a branch of mathematics which describes the way in which the proportions of self-similar shapes remain constant at different orders of scale.

Framingham Heart Study a long-term medical research project which began in 1948 in Framingham, Massachusetts, USA. It followed several generations of residents to identify heart disease risk factors, such as high blood pressure, high cholesterol, smoking, and lack of exercise.

Framingham Risk Score a tool for predicting the likelihood of cardiovascular disease at different stages of life which is based on a set of easily measured risk factors.

F-statistic the ratio of the variance between a group of treatments to the variance within a group of treatments as determined from the mean square values for each condition.

Funnel plot a scatterplot to visually detect the presence of publication bias or heterogeneity according to the distribution of standard errors and overall confidence limits derived from individual studies in a meta-analysis. In this way, the funnel plot provides a measure of study precision.

G

General Data Protection Regulations a set of government regulations which control how personal data is obtained, used, and stored.

Genome-wide association studies (GWAS) a genetic research technique which searches for association between specific genetic markers and particular diseases.

Golden ratio a ratio between two measurements which equals 1.618. It is often linked in geometry and biology to proportionate size changes in fractal structures (the regular spiral shape of a snail shell, for example).

H

Harmonic proportion a mathematical term which explains the relationship between size and shape of objects.

Homeopathy a system of alternative medicine which relies upon highly diluted formulations of natural compounds to treat the progression of an illness.

Hub a node which links one region of the interactome with another.

I

Independent variable a condition which is manipulated by the experimenter.

Inductive reasoning a method of scientific logic which tests an idea by creating a body of observations of a sampled population, which are then used to draw conclusions about the population as a whole.

Inference the process of drawing a conclusion from a set of premises. Logical reasoning uses inference to link premises to conclusions.

Interactome the network of interactions between biomolecules in a cell or organism.

Inter-study heterogeneity a measure of variation in results between studies that have tested the same, or similar, experimental hypotheses, in comparison with random variation.

Isometric the proportions of a shape which remain constant at different orders of size magnitude.

K

Kolmogorov–Smirnov test an alternative normality test which is used to determine if two samples come from different populations. It is often used to assess normality in large or non-parametric data sets.

L

Latin square an experimental array in which treatments are arranged so that they appear exactly one time in each row and each column of a grid.

Least squares regression a line of best fit through the data in a linear model which can be used to predict the behaviour of dependent variables.

Linear modelling any statistical model which assumes that the relationship between the variables is linear.

Logical statement relates to logic which is the study of reasoning and argument. A logical statement is either true or false but not both.

Longitudinal study a research method that involves collecting data on the same set of variables and from the same sample group over time, often years or decades.

M

Machine learning a process using algorithms, statistical models and logic that enables computer systems to operate without following instructions from a human operator.

Mann–Whitney U Test a non-parametric test that is used to compare two sample means from the same population.

Margin of error a standardized measure of random sampling error in a data distribution. It is often used to communicate tolerance limits for experimental error in a system under study.

Massive data mining a process of extracting information from large data sets using a combination of artificial intelligence, statistics, and multi-national database systems.

Maximal statistical model a conservative statistical model which allows the researcher to retain all assumptions that might affect the outcome of an analysis.

Mean a measure of central tendency given by the average value in a data set.

Mean square (MS) sum of the squared distance of each data point to the mean.

Median a measure of central tendency given by the middle number in a data set.

Meta-analysis the statistical combination of results from several studies which aims to increase the precision of measurement of an outcome.

Minimum Adequate Model (MAM) a statistical model which has been refined to include the minimum number of predicting variables for a specific test.

Mode a measure of central tendency given by the most frequently occurring number in a data set.

Molecular chaperone a protein which prevents damage, misfolding, aggregation, and degradation of other protein molecules during periods of cell stress, caused by heat, for example.

Morphological warping the study of how varied anatomical shapes arise from a common basic structure.

N

Natural selection the process whereby organisms better adapted to their environment tend to survive and produce more offspring.

Network motifs repeated interactions which define the distinct characteristics of part of the interactome.

Node an interacting point between several biomolecules in the interactome.

Noise the variation within sample groups.

Non-parametric data which are not defined by fixed parameters and so do not conform to a defined data distribution.

Normal distribution a graphical presentation of data where the frequency of values is distributed symmetrically around a central measure such as the mean, median, or mode. Most frequent values fall around the mid-point and least frequent at either extreme of the distribution. The spread of data in this distribution takes the shape of a bell-shaped curve.

Null hypothesis the prediction of no effect or outcome from an experimental manipulation.

O

Observation a process of information gathering using the primary human senses of sight, sound, touch, taste or smell.

Odds ratio the ratio of the probability of the event happening in an exposed group versus a non-exposed group.

Office for National Statistics the largest independent producer of official statistics in the United Kingdom.

Omnibus test a statistical test to determine if differences exist between multiple groups in an analysis. It normally precedes a post hoc test which assigns probability to each test group.

One-tailed probability distribution a probability distribution within a statistical table which allows for the possibility that a measured outcome will fall only at one end of the normal distribution.

Outliers data values which are unusually distant from the spread of other data points in a distribution.

P

Paired test a statistical test which looks for treatment effects in two related sample groups from a population.

Parametric data which are defined by a set of fixed parameters. These define the shape of the data distribution.

Pearson's correlation coefficient (r) a method for measuring the strength and direction of the relationship between two continuous variables.

Peer review a process of oversight in which the standard of research reported in a scientific study is evaluated by other scientists with expertise in the same field of work.

Percentile a set of ranges obtained by subdividing a data distribution into 100 intervals.

Permutational multivariate analysis of variance (PERMANOVA) a non-parametric multivariate statistical test which makes comparisons between multiple groups using geometric measures of variance.

Phenotypically pleiotropic a biological effect where a single genetic mutation or protein defect can produce multiple changes to the function of cells and organ systems.

Placebo a blank treatment that looks real but which does not contain the active component that the scientist wishes to test. Researchers use placebos in experiments to compare how people respond to a real treatment versus one that has no effect.

Population a defined collection of items with common characteristics which make them distinct from other items around them. Biologists usually think of 'items' as living things but in statistics, an item might be anything, from a series of similar events to molecules or objects.

Post hoc test a statistical test that is conducted after a study has been analysed using an omnibus test and assumptions about normality have been checked. It is used to assign probability values to test group comparisons.

Preferred Reporting Items from Systematic reviews and Meta-Analyses (PRISMA) a minimum set of quality checks which help scientific authors to report how they collected and assessed literature that is to be included in a systematic review.

Premise a statement or idea which serves as the basis for a logical line of reasoning. In science, a law can serve as a premise from which we might predict the outcome of an experimental test.

Principal component a new variable derived from linear combinations of the original variables in a multi-component analysis.

Principal Component Analysis (PCA) a statistical method for identifying the essential features within a multi-component analysis which best explains the variance amongst all original variables.

Probability density the probability of a selected data point lying within a defined range of values in a data distribution.

Q

Quantile–Quantile (QQ) plot a plot of quantiles of one data set against another which is used to determine if two data sets share the same distribution pattern.

Quartile a set of ranges obtained by subdividing a data distribution into four intervals.

Qubit a quantum version of a computer 'bit', which can occupy more than one quantum state. This ability makes quantum computers far more powerful than regular computers for complex functions.

R

Random sampling method of sampling in which each member of a population has an equal likelihood (probability) of being selected.

Regression line a line of best fit which describes the relationship between the dependent variables on the y axis, and independent variables on the x axis, on a cartesian (x–y) graph.

Relative risk the ratio of the probability of an event occurring with an exposure versus the probability of the event occurring without the exposure.

Risk the probability of a given outcome measured as the ratio of the chance of a possible outcome divided by the number of all possible outcomes.

Risk factor a biological change which is caused by an event, and which alters the course of its outcome.

Risk marker a biological change which is caused by an event, but which does not change its outcome.

Risk of error the likelihood (probability) of drawing an incorrect conclusion when testing a hypothesis.

S

Sample a group of randomly selected members of a population which can be used to draw conclusions about the characteristics of the same whole population.

Scaling factor an exponent from a regression through data on a log–log plot which describes allometric relationships.

Scepticism the act of doubting the truth of an assertion about an assumed fact.

Scurvy a disease caused by vitamin C deficiency which presents as weakness, anaemia, immune imbalance, and bleeding of soft tissues, such as the gums in the mouth.

Sexual selection selection arising through preference by one sex for certain characteristics in individuals of the other sex.

Shapiro–Wilks test a statistical test which is used to determine if a data set is normally distributed. It is often used to assess normality in small data sets.

Shortest path an energetically favoured interaction between nodes in the interactome.

Sign Test a non-parametric test which determines if two paired groups are different based on the direction of change between each paired measure.

Signal the distance between two sample means.

Skew the tendency of data points to distribute more often towards one end or the other of a range creating an asymmetric data distribution.

Spearman's rank correlation (r_s) a non-parametric measure of correlation using the ranked order of two variables.

Special-cause variation unexpected changes which are caused by factors that are not related to the system under study.

Square–Cube Law a geometric relationship which predicts that when an object increases in size, its surface area grows with the square of the scaling factor while its volume grows with the cube. As a result, larger objects have much greater volume relative to surface area, which affects strength, heat loss, and biological scaling.

Standard deviation the square root of the variance of a data set. It serves as a convenient measure of data dispersion on a normally distributed data set.

Standard Normal Distribution Table a table which can be used to find the likelihood that a data point might occur at any given place within a normal distribution.

Statistical power a measure of the ability of an experimental design to detect an effect if it is truly present. In this way, power helps the researcher to assess the risk of type II error.

Statistics a branch of mathematics that provides methods for collecting, organizing, and describing numerical observations, and uses probability to identify patterns and relationships within those data.

Stereoisomers two molecules with the same atomic composition which are arranged as mirror images of one another.

Student's t-test a method of hypothesis testing which relies upon the t-statistic to determine if the means of two sample groups are different to one another.

Study precision a measure of how close a statistical measure is to the true value of a population.

Systematic review an objective, reproducible method to find answers to a certain research question, by collecting all available studies related to that question and reviewing and analysing their results.

T

Topological communities groups of nodes with heightened connections to one another.

t-statistic measures the size of the difference between two means relative to the variation within the samples.

Tukey's honestly significant difference (HSD) a type of post hoc test which looks for pairwise differences between groups in ANOVA whilst adjusting for the risk of type I error within each comparison.

Two-tailed probability distribution a probability distribution within a statistical table which allows for the possibility that a measured outcome could fall at either end of the normal distribution.

Type I error rejection of the null hypothesis when it is likely to be true. Often referred to as a false positive result.

Type II error rejection of the test hypothesis when it is likely to be true. Often referred to as a false negative result.

U

Unpaired test a statistical test which compares the means of two unrelated sample groups from a population.

V

Var(line) a measure of how far each variable lies from a line of best fit through the data in a linear model.

Var(mean) a measure of how far each variable lies from the response mean in a linear model.

Variance or deviation a measurement of the spread between numbers in a data set.

W

Western blotting a laboratory technique which is used to detect specific proteins from a cell or tissue sample.

Wilcoxon Signed Rank Test a non-parametric test which compares two paired groups by assigning a rank and direction of change to each paired measure.

Z

Z-score the distance in standard deviations between a data point and the mean in a data distribution.

INDEX

L

M

N

O

P

The manufacturer's authorised representative in the EU for product safety is Oxford University Press España S.A. of el Parque Empresarial San Fernando de Henares, Avenida de Castilla, 2 – 28830 Madrid (www.oup.es/en or product.safety@oup.com). OUP España S.A. also acts as importer into Spain of products made by the manufacturer.

www.ingramcontent.com/pod-product-compliance
Ingram Content Group UK Ltd.
Pitfield, Milton Keynes, MK11 3LW, UK
UKHW062029240626
472590UK00016B/76

9780198862253